ENVIRONMENT AND SOCIETY
An Introductory Analysis

ENVIRONMENT AND SOCIETY
An Introductory Analysis

Brian Harvey
and
John D. Hallett

The MIT Press
Cambridge, Massachusetts

First MIT Press paperback edition 1977

Second printing, 1978

First published in the United Kingdom 1977 by
The Macmillan Press Ltd

Library of Congress catalog card number: 77–75994

ISBN 0 262 58033 0 (paperback)

Printed in the United States of America

Contents

Preface

The environmental debate—a combination, difficult to define, of such overlapping issues as population growth, pollution, resource depletion, the distribution of resources between developed and less-developed nations, the relationship between economic growth and social welfare, the pace and direction of technological change—began to take shape during the late 1960s. Of course the separate issues were there before, and some individual writers had begun to make connections, but more widespread awareness stems from this period.

Books of the period have dramatic titles such as *Planet in Peril*, *Only One Earth*, *The Closing Circle*, *A Blueprint for Survival*, and *The Limits to Growth*. Yet it was in the late 1960s that the developed nations of Japan, Western Europe and the United States of America were experiencing unprecedented levels of material prosperity. There was confidence that, in the First Development Decade (as the 1960s were designated), the transfer of capital and expertise would also set the less-developed countries on the path of self-sustaining growth. Later, a recession in world trade and in economic growth, and more particularly the oil crisis (although, arguably, specific phenomena not centrally relevant to the general environment issue), did yield some insights into the practical implications of the Club of Rome's resource-depletion calculations. Initially, however, widespread scepticism was understandable, given the economic background, the dramatic (sometimes melodramatic) presentation of environmental issues, the speculative nature of the topic and the range of emphases and opinions.

Whatever the swings of magnitude and focus of public and official interest, the authors of this book accept that the environment issue is here to stay and will become increasingly integrated into the stream of public policy. This is reflected partly in the growth of undergraduate courses in this area, which the book aims to serve. It therefore sets out to summarise the environment debate and survey the issues that compose it, from both a natural and social scientific viewpoint. Primarily intended for students on multidisciplinary courses, it:

(1) plots the development of the environment issue;

(2) outlines the different positions that have been taken on the nature, causes and treatment of environmental problems;

(3) presents the readily available overview of the different elements of the topic, in particular: (a) the basic natural science relevant to ecology, resources, pollution and population growth; and (b) the politics and economics of environmental problems and policies.

The book is written with the non-specialist reader in mind. Coming to a broadly based course on the environment, someone with a natural science background will find here a brief summary of relevant scientific topics. He will also find an introduction to the problem in its social–political–economic context, asking why environmental problems occur and what responses might be made to them. The same applies to a student with a social science background. The essential material for the non-scientist on ecology, pollution, etc., is provided in the early chapters. Contact with environmental issues on earlier courses will probably have been as brief illustrations of particular topics like external costs in economics, or pressure groups in political theory. So in the later chapters there will be found a systematic application of the social scientific approach to environmental issues. Suggestions for further reading on specific topics in each chapter, and detailed references, are grouped at the end of the book.

Chapter 1, 'Basic principles of Ecology', establishes the basic idea of global ecology by examining natural processes that occur in the absence of man: the conservation laws for materials and energy; cycles of materials and flow of energy as they affect life; the efficiency of energy transformations; photosynthesis, food chains and food webs. The factual information available on population globally and in particular countries is presented in chapter 2, together with the arithmetic of population growth. The ideas of exponential growth and doubling period lead to a discussion of trends and forecasts, the factors that control the growth of animal populations and man's ability to avoid these natural controls. The chapter then considers the basic survival needs, the production of food and the techniques of agriculture. Chapter 3 is concerned with resources (the technical and economic aspects of meeting demands for fuels and materials; the energy budget and equilibrium sources of energy; the lifetimes of scarce resources) and pollution (the disposal of energy and materials as waste; dilution and mixing in water and atmosphere; the disturbance of natural cycles).

Chapter 4 surveys the development of the environment debate and summarises the positions taken on its causes and treatment by some of the main contributors. The focus of chapter 5 is on the framework of society within which environmental problems occur. The decision-making processes of society are reviewed and the point made that environmental problems are better regarded as an integral part of ongoing political–economic processes rather than as a separate issue. On the basis of this analysis, chapter 6 considers various approaches to environmental policy: taxes, subsidies and the role of government and law in the fields of pollution, resource depletion, population growth, the development of technology and the social–industrial environment.

The final chapter asks the question: will environmental problems respond to these reforms, or do they demand more fundamental changes in the structure of society? By referring to writers such as Illich, Ellul, Roszak and others, it examines the questions that command attention today. What are the forces that are thought to shape societies? What possible futures are projected or foreseen for ours? Is a post-industrial society evolving? If it is, what form would it take, and would continued economic growth be a part of it? What will be the position of technology: a tyranny of high technology or a humanised intermediate technology? What path will we take to any future society—can we build on ideas and institutions already in existence, however radical, or will there need to be a catastrophe?

CHAPTER 1

Basic Ecological Principles

The beautiful photographs brought back by the early astronauts were historically important because, for the first time, men had been able to stand far enough away from their departure point to be able to see the whole Earth spinning in space. This new view of Earth corresponded in time with a growing awareness that a new outlook on the planet as a whole, a global perspective, was needed if the delicately balanced conditions that support life as we know it were to be better understood. Ecology is itself a new word, coined only 100 years ago and in general use only over the past decade, and global ecology is very much a new field of study. The word ecology means literally the study of living things 'at home' and its objective is to discover and understand the relationships which exist between living things and their environment. The early ecologists worked on situations where the local conditions made it possible to study a small environment, a pond or wood or even a rock crevice. They realised that the life, even within these restricted zones, was affected by the wider surroundings but they put limiting boundaries around their study to reduce the factors involved to a manageable number. Drawing boundaries is artificial; the wind and rain sweep through the wood, animals and birds come and go across these boundaries. One very special recent study of an unusual environment is that of the astronauts in their command module. Here the boundaries could be defined with much greater validity than in most situations on Earth and here an understanding of the environment was vital. Apart from solar energy and radio waves almost nothing crossed the boundaries of their system and the necessity to take and maintain all that was needed to support life, (as well as devising a system to cope with the waste materials) was obvious. For us the need is also clear; we need an understanding of the relationship between the crew of spaceship Earth and the systems of our spaceship, which have enabled life to survive on this rather unusual planet for millions of years, if we are to ensure the maintenance of these systems for future generations.

1

Such a global approach to ecology will have its critics. The vastness of the field of study and its complexity might make it virtually impossible to get any useful answers. It is significant, however, that in many respects as we move to a larger scale of study, patterns and relationships often become more, rather than less, obvious. An insurance company builds its business not on the behaviour of this or that individual but on the overall statistical picture of the lives of large numbers of people. The statistical data are gathered from individual case records; the trends emerge when these are put together. When we want to see the pattern of a wood we must stand far away from the individual trees and, similarly, we can predict the behaviour of a population more successfully than we can that of an individual. It will also be argued that global ecologists have already come up with answers that are clearly seen to be wrong because their information or their assumptions were not good enough. This has already been shown to be true, even in the brief history of the science, but the criticisms should not be taken to be an argument for less global ecology. There is a necessity in many fields of human activity to do rough sums, even with few data and much uncertainty about the importance of factors affecting the answers, because we need rough answers to make immediate decisions. Computers can help us handle the vast amounts of data that are available and make it possible to try out suggested possible relationships between such factors as population, resource usage and food production. Every attempt to obtain even rough answers can provide some insight into the characteristics of a better method of approach. The purpose of this chapter is to attempt to identify the processes which go on in the ecosystem and to establish the general principles of ecology in simple terms.

1.1 The Movement of Materials in the Biosphere

The best starting point is probably that offered by the American scientist, Garrett Hardin, when he used the phrase 'We can never do merely one thing!' We regularly behave as though we are doing merely one thing; we see ourselves as consumers when in fact what we are doing is using materials and energy at a particular stage in a longer process. We read this page—on paper which was once forest, trees which used, for their growth, materials available in the air, the rain and the earth, and energy from the Sun. The page will sooner or later go, forgotten, into the dustbin, to waste disposal, perhaps by burning in which case the ash will be scattered and find its way back into the air, the rain and the earth. The materials have never been consumed; they remain, somewhere, in different forms. Behind this lies the fundamental idea of the conservation of matter, *that the basic raw materials are not created or destroyed in these processes and that accounting for them is a worthwhile activity.*

The staff of a bank expect to be able to account for the money handled during a day even if that money has been transferred into or out of the building, or stored in it, as coins, notes, bankers orders, or cheques, and so on. No bank manager will accept happily that money is unaccounted for, because he treats money as being conserved; ecologists, similarly, see it as their business to account for the basic raw materials like carbon, oxygen and nitrogen. In our accounting we shall have to recognise that these materials may combine together and appear in different forms. Thus carbon can combine with oxygen to form carbon dioxide gas and our attempts to trace the flow of carbon must include any in this form. The term 'basic raw materials' must be clarified since a distinction between carbon and oxygen, taken separately, and carbon dioxide, has been implied. Carbon dioxide can be broken down into its two components, carbon and oxygen, which in almost all the processes known to man, cannot be broken down further or synthesised out of even more basic materials. For this reason they are known as elements. The exceptions lie in the group of processes known as nuclear (or sometimes atomic) reactions, which, though of great importance, do not affect the general line of argument. It is also the case that, even in the study of nuclear reactions, an accounting process is used successfully and it is based on the principle of conservation in a more subtle form.

Our Earth became a recognisable planet something like four or five thousand million years ago and during the early stages of its formation probably lost some of the lightest gaseous materials from its atmosphere. By the time the atmosphere took on the established form we now know and life appeared this process of loss was virtually completed. There has been no influx of material from outside the atmosphere on any significant scale since and so it is possible to consider the Earth (or, rather, the materials we find on it now) as a closed ecosystem. Imagine a spherical boundary around the Earth and its atmosphere and consider an element like carbon. No carbon can be detected crossing this boundary coming in or going out, and since carbon is indestructible the total amount of carbon must remain constant. This ecological principle can be stated simply in the form – *the resources of the Earth are finite, limited in quantity.* There is no more carbon available today than there was when man first appeared. The carbon that is here, in the system, is on the move combining with other elements, separating and recombining in various processes. What is true for carbon is also true for oxygen, lead and mercury, or any other element we may consider. The bulk of the material of the earth is not available to man, however, and when we are considering the resources of the Earth we must remember that very rarely do we extract materials from depths greater than 2 km which is merely scratching the crust of a sphere of radius 6400 km.

If we consider any one of the elements present on Earth and the processes it undergoes we find that the sequence is repeated and the same molecules of the element go through the same sequence of processes repeatedly; an understanding of the nature of these cycles and the rate at which an element passes through its cycle is at the heart of global ecology. We find that some of these cycles involve living organisms and so are part of the process of life while others, usually the slower cycles, can be completed with few biological links. We shall consider in this chapter three contrasting cycles as examples to illustrate the wide range that exists and to enable us to discuss the details of particular real situations.

1.1.1 *The Water Cycle*

The bulk of the water resources of the Earth—roughly 97 per cent in fact—is in the oceans but at any instant some of the water is moving through a sequence of events known as the water cycle. This is shown in simple form in figure 1(A), and 1(B) shows the quantities involved in the form of a flow diagram. Taking the simpler picture first (1A) we see that the sun's energy falling on the sea warms it and evaporates some of the water at the surface. This warm vapour rises and forms clouds. Some condenses again to form large enough droplets to fall as rain on the sea but other clouds are carried away by the wind over land where they fall as rain. Taking an overall picture both on a global scale and smoothing out seasonal variations we find that quite a large part of the water which falls on the land stays sufficiently near the surface long enough for evaporation to take place and new clouds to form. Some of the water is used by green plants for growth, some seeps underground but most runs off in streams and rivers to return to the sea. More rain falls on the land than is evaporated from it in any given period and the balance is maintained by a supply of windborne clouds from the sea and by runoff water in rivers. The energy supplied by the sun keeps the cycle flowing.

In figure 1(B) we see the quantities involved. The flow figures are given in km^3/day (1 km^3 = 220 000 million gallons) and the figures in brackets are the amounts estimated to be stored at any one time in thousands of cubic kilometres. The figures are enormous and defy the imagination but as all the data are expressed in the same units comparisons can be made. One obvious point is the small amount of water vapour in the atmosphere at any time; if we consider the actual flow rate for the whole cycle as 120 km^3/day then the atmospheric reservoir holds only 4 months flow. This leads to the idea that

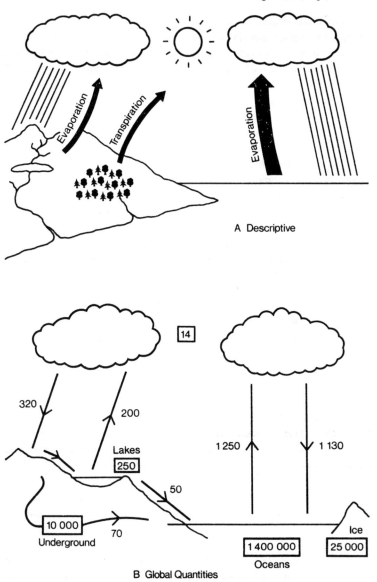

Figure 1 The Water Cycle. Quantities in thousands of cubic kilometres for stored water (figures in boxes) and in km³/day for flow. (Data derived from the Report of the Hydrological Decade Conference[1])

the cycle must be quite rapid and must be measured in months rather than years. The average stay in freshwater lakes and rivers is only about 16 months and in underground waters about 15 years. The water used by man comes from the 50 km^3/day runoff water in rivers, streams and the 70 km^3/day flow through natural underground reservoirs. At present, man tends to meet his needs by drawing on those surface waters which he can store in artificial reservoirs or by drawing directly from rivers. Evaporation from reservoir surfaces is the cause of quite a considerable loss; in Britain on a warm day, four or five litres may be lost from each square metre. It is also true that the rivers which feed the reservoirs are part of the transportation process in sedimentation cycles (see p. 8) and so silting up will be a continual problem. In some places water is trapped underground in natural reservoirs (known as aquifers) and can be drawn to the surface by pumping. While evaporation and silting are not problems in such cases, excessive removal of water can lead to harmful effects. These include lowering the local level of standing water below the surface of the soil so decreasing the amount available for uptake by the roots of crops. The effect of man's activities on the water cycle is discussed in chapter 3.

1.1.2 *The Carbon Cycle*

The carbon cycle and the water cycle have been chosen as the first two examples because they deal with the fundamental materials of life and interlock together. In our preliminary look at the water cycle we assumed that all the water stayed as water throughout and was not involved in chemical reactions. The reference to the water uptake by the roots of crops, however, leads to the consideration of the way some of the water becomes a part of living plants. This consideration not only involves water but also the carbon (and oxygen) with which it is linked chemically to form the more complicated molecules which are the building materials for plant and animal tissue. Green plants are able to use the energy of the Sun to absorb carbon dioxide from the atmosphere and to combine the carbon and oxygen atoms with the hydrogen from the water to form large carbohydrate molecules. The energy from the Sun is needed to perform the rearrangement of the simple molecules to form the chemical bonds between the atoms in the larger molecules. This is the process of photosynthesis; it is carried out in the presence of carbon dioxide and water and light and it results in the production of free oxygen (from the water) which goes back into the atmosphere. Some of the 100 000 million tonnes of plant material produced by photosynthesis on the Earth's surface each year is eaten by animals and the carbon moves further along the cycle. Both plants and animals use carbohydrates as fuels in respiration in which the large molecules are broken down by what is, in effect, a very slow burning process, which releases water and carbon dioxide back into the atmosphere. Not all the living material is broken down in this way; some dies and decays and again carbon dioxide is released as micro-organisms of decay break down the material into its simpler components. Thus the carbon cycle must take account of the appearance of the carbon in gaseous form as carbon dioxide (CO_2) and as part of the carbohydrate material of plants and animals. The cycle over land is shown in figure 2 in simple form. Over the sea we have

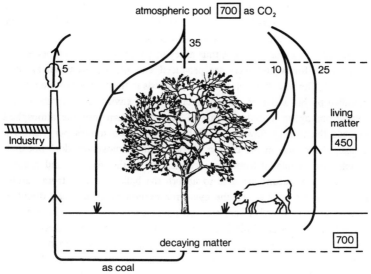

Figure 2   The Carbon Cycle over Land. Quantities in thousand million tonnes for stored carbon (figures in boxes) and in thousand million tonnes per year for flow. (Data from B. Bolin, *Scientific American*, **223**, no. 3.)[2]

an equivalent process in which carbon dioxide is dissolved in seawater and is available to single-celled organisms in the upper layers of the sea (phytoplankton), which can use the sunlight for photosynthesis to form carbohydrates. This material becomes the food for fish, which can use the dissolved oxygen which is given out in the photosynthesis process for respiration, or it decays and falls to the sea bed. Thus, there are similarities in both the land and the sea-based parts of the carbon cycle particularly in the respect that carbon may be, for part of its cycle, locked in the material of plants and animals.

Almost all these materials are able to decay so giving up their carbon hydrogen and oxygen in this relatively simple and direct manner. A very small quantity, however, is buried in particular climatic and geological conditions such that decay is halted at an early stage. Over millions of years this material may be fossilised and compressed to form a store of carbon (and energy). It is from this source that our fossil fuels, oil, coal, etc. are derived. In the absence of man these stores would remain unused under the surface, perhaps rising when large earth movements occurred, but still relatively stable and, for the time being, out of circulation. One of man's most significant activities has been the exploitation of these rich pockets of stored energy; the comparatively rapid burning of the fossil fuels over the past hundred years has resulted in the release to the atmosphere of carbon dioxide at a rate which is more rapid than at any other stage in the cycle and more rapid than the rate at which replacement stocks of carbon fossil fuels could be produced. The carbon cycle is thus a key to understanding the processes by which both foods and fuels are produced and will be discussed again in chapter 3. The rate at which the cycle proceeds in the absence of man, and is modified by his activities, are key factors in the environmental debate. It is clear from figure 2 that the rate of flow of carbon into the atmosphere exceeds the rate of flow out of it by 5000 million tonnes per year. The measured increase in the carbon dioxide in the atmosphere does not quite match that calculated from the excess shown in the diagram and it seems likely that the oceans are able to absorb the bulk of output from burning fossil fuels. In addition to the cycle for carbon that has been discussed already there is a further very slow burning carbon cycle which does not involve living materials. This is a geological rather than a biological process and involves the formation and erosion of carbonate rocks in the Earth's crust. The cycle is typical of those discussed in the next section of this chapter.

1.1.3   *The Mercury Cycle*

The water and carbon dioxide cycles, because they involve materials in liquid or gaseous forms, are comparatively rapid cycles. Even when they combine together to form the materials of plants and animals the residence time, even when the decay process is considered too, is short by comparison with the times involved in the cycles of materials which make up the Earth's crust. These cycles follow a general pattern of erosion, sedimentation and compression followed by more violent processes which bring the rocks to the surface again and expose them to erosion. They have, until recent years, been the subject of study by geologists and of little concern to ecologists because they take place so very slowly and, except where earthquakes and volcanoes cause damage and loss of life, seem to have little impact on man's survival. It has become evident in recent years that some of the materials which follow these geological cycles can, as a result of geological accidents or the activities of man, be gathered into local concentrations; they can combine chemically with the molecules of living materials and in living tissue they can be extremely toxic, even in small concentrations. Mercury is an example of this kind of material. Rocks containing mercuric ores are eroded (or mercury is ejected into the atmosphere by volcanoes) and in a highly dispersed state goes through the sedimentation, compression folding processes. Despite any suggestions derived from the popular name for mercury (quicksilver) this cycle is very slow and tends to disperse rather than concentrate the mercury. Mercury is, however, used in industrial processes; the production of paper and pulp, caustic soda, chlorine and fungicides all use mercury either as a prime ingredient or as a processing agent. Mercury has been concentrated at such industrial plants and some has been lost, or even discharged into the rivers and coastal waters nearby. As a simple metallic element mercury is not a widespread danger to man; it can, however, be converted into highly toxic organic forms by certain aquatic micro-organisms producing the methyl- and alkyl-mercury compounds. These, in turn, are concentrated in the tissue of fish (which feed on the micro-organisms) and, when the fish is eaten, find their way into the tissue of man with disastrous results. There is still uncertainty about the general background level of mercury and about the danger level in human tissues but in the light of local catastrophies which have been widely publicised steps have been taken to control the use and release of mercury and to monitor levels particularly in fish and other seafoods. Mercury has been chosen as an example but very similar situations exist where other heavy metals, such as lead, cadmium, chromium, arsenic and nickel are concerned. All are toxic and persistent; all are in increasing use in industrial processes.

Between the stages of erosion and sedimentation the transportation of the finely powdered products of erosion must take place. Some of this transportation takes place in the atmosphere, some in the water which runs off the land. Some airborne dusts are carried for long periods and deposited far away from their point of origin; some at high altitudes may circle the Earth several times. Into the slower geological production of dust there are also more violent intrusions. Some arise from volcanic eruptions and similar events but some are the products of man's activities. Good fertile land can become a desert when the crops which are planted do not bind the topsoil as the original natural vegetation did, and the removal of the all-plant crops at harvest time leaves the topsoil exposed. More violent intrusions result from nuclear explosions with the additional hazard that the intruding materials can be radioactive and highly toxic for long periods. Waterborne material has an obvious link with growing plants which may be used as foodstuffs and an even more direct link with drinking water supplies but is considerably easier to monitor and control.

In summary it would seem that natural geological cycles pose only rare local problems, though these may be severe, rather than global problems. The processes of mining, industrial production and disposal of waste can be seen as modifying both the speed at which processes occur within the cycle and the global distribution of the materials by participating in the erosion process and modifying the transportation–sedimentation stages.

The three cycles considered in detail are examples chosen to illustrate the general ideas involved. A more complete study would examine the oxygen cycle and the nitrogen cycle which both involve biological processes and the phosphorus or sulphur cycles which while providing elements essential to life are largely geological and slow moving. Figure 3 shows the general pattern of these cycles which are sometimes spoken of an biogeochemical cycles to indicate that materials pass through stages that involve chemical combination sometimes within the material of living things and sometimes in geological processes. It shows also that there can be cycles within cycles and that some are so slow moving that we can describe them as losses in the sense that, for example, deep ocean sediments may reside for millions of years without disturbance and, in this sense, out of the main stream of circulation.

1.2   The Flow of Energy in the Biosphere

We have discussed the cycles in which materials circulate and in each case it has been clear that another vital resource has been used—energy. An understanding of the nature and role of energy is the key to understanding any process of change. Energy is not easy to define. A famous physicist called it 'the go of things' which, though not an elegant definition, conveys well the most distinctive characteristic. A more prosaic but useful description is the capacity to lift a weight. Energy may appear in many forms. The water at the top of a waterfall has the capacity to break stones at the bottom—or to drive turbogenerators and supply electrical energy. At the top it has stored (potential) energy, associated with its height, as it falls this energy is transformed into the energy of movement (kinetic energy) which is transferred to the rotating parts of the generator for conversion to electrical energy. This electrical energy can lift a weight or, using an electric heater, boil some water. The energy which is supplied in electrical form finally reappears as heat energy. Energy is stored chemically in fuels, which when burnt, can give out heat, and in engines can again be used to lift a weight. Just as we are able to

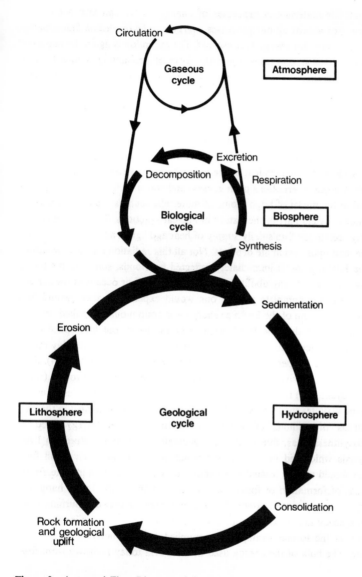

Figure 3 A general Flow Diagram of the Relationship Between the Cycles Discussed in Chapter 1

account for materials in processes of change so we can also account for energy. For almost all our processes *energy can be treated as indestructible and so we say that energy is conserved.* The exception is again the important group of nuclear processes where the form of accounting is based on the more comprehensive conservation principle that treats mass and energy together in order to allow for the transformation of mass into energy.

Unlike the basic materials of the ecosystem, however, the total stock of the Earth's energy is not fixed but is continuously supplemented by the daily income from the sun and depleted by radiation out into space.

There is a capital stock of energy laid down during the process of formation of the Earth and evident in the heat which flows from the interior of the Earth in volcanic eruptions or in geysers and warm springs. Energy was also stored in the nuclei of large atoms of materials, such as uranium, which can be made to yield part of their store in nuclear reactions. The income is most clearly seen in the Sun's daily supply of heat and light, radiated across empty space, and upon which life depends. Not all this radiation reaches the surface of the Earth, some is immediately reflected by clouds, some warms the atmosphere, leaving only about half to warm the land or oceans at the surface. If this was the complete picture one would expect that there would be a general heating up of the Earth as energy was continuously supplied. What in fact happens is that the Earth itself radiates energy out into space and a dynamic balance occurs so that locally heated regions during the day radiate faster and the temperature of the Earth stays the same (see figure 4). Solar energy is thus elusive and better thought of as part of a flow process (rather than a cyclic one) from which we can only benefit in passing unless it can be in some way, trapped and stored. This does happen; it happened in the past, hundreds of millions of years ago when the solar energy fixed by photosynthesis was, due to unusual geological conditions, stored in living materials which did not decay and which now constitute our fossil fuels. These should also be treated as a capital stock rather than an income, for the process of formation of fossil fuels is not continuous and is not happening now to any significant extent. The major difference between burning wood from a newly fallen tree and burning coal lies in the time scale involved. The energy in the former example was supplied only a matter of a few years before. The bulk of the energy used in the world today is drawn from fossil fuels.

The process of photosynthesis still goes on but the solar energy fixed in that way is either reradiated when the plant material decays or is used as food by an animal. The saying 'All flesh is grass' is, in a real sense, true and a key to understanding; animals need carbohydrates both for building up their own body materials (for growth and replacement) and for fuel to provide energy for activity, and derive both from their food. This food may be plant material eaten direct or animal flesh which has itself at an earlier stage been built up from the material and the stored energy of green plants. When an animal uses food to provide energy for warmth or activity it is carrying out a process chemically very similar to burning (slow burning at low temperature) which breaks down the molecules and releases the energy.

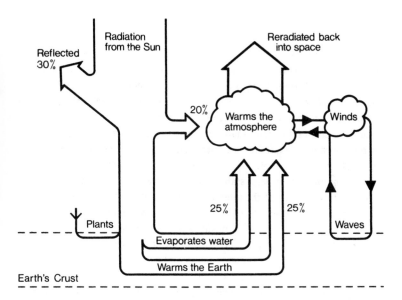

Figure 4   The Budget of the Earth's Income of Solar Radiation

1.3   Interrelationship between Materials and Energy

So the cycles of materials and the flow of energy are intimately related. Cycles of different materials interlock with each other at different stages as molecules are built up and broken down. In the process of photosynthesis we have the release of oxygen; this in itself is a process vital to life because animals need oxygen in order to break down the plant materials which they have eaten and they need it in the form of a gas which they can breathe. When it has played its part in the lungs and bloodstream of the animal it is rejected as waste because it has combined with carbon to form carbon dioxide, useless to the animal, vital to the plants. The cycles, together, form a web in which a slight disturbance in any process can have repercussions throughout several different material or energy cycles. Felling trees, which may have taken thirty years to grow, to make paper is at the simplest level transferring carbohydrate molecules to an extra loop in their cycles. But the removal of the trees changes the hillside environment from which they came; the shade has gone, rain water runs off the hillside more rapidly, small animals and birds move away and insect populations may rise unchecked by natural predators. To provide one edition of a major national newspaper costs us several acres of forest.

We can picture the cycles of materials and energy which had been established gradually over the millions of years which elapsed before man appeared on the scene, cycles which were woven into a complex web which was resilient to disturbance. This appears to be a further basic principle of ecology—*that natural complex systems are generally stable and able to return to their equilibrium state after disturbance.* The tropical rain forest will be able to endure all but man's intervention but in contrast a field of wheat is in constant need of man's care and protection if it is to survive even the normal range of variation in climate and similar natural disturbances. The wheatfield is an example of a simple system—a monoculture—which, lacking complexity, is less stable. Long before men were aware of this ecological principle or of the interdependence of the natural cycles in a web they began to intervene and modify. Sometimes by cutting and burning the timing of cycles was altered, sometimes by technological activity materials were redistributed geographically. Almost all of this activity of intervention has been and still is carried out in accidental or deliberate ignorance of the ecological consequences. In ecological terms man must be seen as part of, rather than apart from, these cycles, dependent on them for survival yet able to modify them from within.

When we consider the flow of energy in foodstuffs a further ecological principle emerges. We speak of food chains in which the green plant is seen as the primary producer and the herbivore (plant-eating animal) occupying the first level in the chain as primary consumer. The flesh of this primary consumer can be food for a flesh eating animal (carnivore) at the second level and in many instances we can identify a third and fourth level consumer. Insects may feed on riverside plants, small fish feed on the insects only to be in turn eaten by a bird or perhaps man. Man is normally at the end of the chain although on a global scale he is most often a herbivore primary consumer. As we try to account for the quantities of energy involved at each stage we find that all transfers involve loss of energy. *The ecological principle involved here is that in natural processes, when energy is transferred some loss always occurs.* The efficiency of the transfer of energy in a food chain is very low indeed; rarely is more than 10 per cent of the energy fixed by photosynthesis in plant material actually available to the consumer in the next level as usable energy.

It is interesting to do an energy account for a field of some green plant intended for food. Over a full year only about 10 per cent of the solar energy which falls on the field actually falls on the green material in which alone photosynthesis can take place. Only about 10 per cent of this is actually stored by photosynthesis and about half of what is stored in this way is used by the plant itself for its own growth and development. A cloud of locusts could strip this field efficiently but in doing so they will only receive something like 0.25 per cent, 1 part in 400 of the original solar energy. Man, with selective harvesting and cooking rarely utilises more than 1 part in 1000. This leads to the ecological principle that *for every kilogram of herbivorous animal there must be a very much greater mass of green food plant to sustain it with energy.* When we consider a second or third level consumer we allow for about 10 per cent efficiency in each transfer of energy. The mass of plant material needed to sustain life will be more than 1000 times greater than the mass of consumers it sustains. It is clear that the flow of energy through a food chain is characterised by large losses during transmission. To speak of losses may be misleading here: the energy which does not flow through to the consumer is not lost (in the sense that we cannot account for it) but lost only in the sense that it is not available to the consumer. It ends up as heat energy—but at low temperature and low temperature heat cannot be used to lift a weight or form chemical bonds. The simplest analogy is probably that of hydroelectric power; a small lake high in a mountain can be used to drive turbines and generate electricity. The same mass of water in the valley cannot be used because it is at the lower level and has nowhere to flow downhill to. All processes which occur naturally, in which energy is transferred and utilised, are subject to this basic law, which is called by scientists the second law of thermodynamics. This law can be stated in a variety of ways but for our purposes it has the implications *that energy transfer is never 100 per cent efficient and that the 'lost' energy appears as waste heat destined only to raise the temperature of the surroundings and unavailable for any other use.* Sometimes we can get good value out of fuel. When we use steam to drive turbines (40 per cent efficiency) which drive generators and produce electrical energy (95 per cent efficiency) these machines get hot and this heat energy would raise the temperature of the machines to dangerous levels if they were not cooled. The cooling water receives the energy and could be used to warm houses.

If we consider the whole process of extracting energy from the fuel used to make the steam right through to the electrical energy produced by the generators we must combine the efficiencies of all the stages in the transformation. The boiler efficiency will typically be about 88 per cent (fuel energy to steam), the turbine efficiency 40 per cent (steam to rotational kinetic energy) and the generator efficiency 95 per cent (kinetic to electrical energy) yielding (95 × 40 × 88) per cent, or 33 per cent. In plain words one third of the energy in the fuel goes to producing electricity and two thirds to heat. This heat is used in some situations to warm buildings but, in most cases it is allowed to go to waste, raising the temperature of the surroundings to no useful purpose.

Summary

We can summarise the ecological principles discussed in this chapter in this way. Any process we study is part of a longer sequence of processes and is frequently a stage in a cyclic path through which the basic raw materials of the earth are flowing continuously. The materials, and in particular the elements, are conserved in all but nuclear processes and so we can account for them. There is no appreciable flow of materials across the outer layers of the Earth's atmosphere, in or out, so we must treat the Earth's resources as finite. Energy is supplied daily from the Sun and is involved in all processes of change but man is currently using, for most of his needs, energy from fossil fuels laid down millions of years ago. The cycles of materials and the associated flow of energy are all interrelated in such a way that waste from one is a raw material for the next. In naturally evolved systems animals eat a variety of foodstuffs and are preyed on by a variety of predators; such a system is complex but resilient and capable of restoring balanced conditions after a disturbance. Man has set up monocultures which lacking complexity are precarious and need protection. In the transmission of energy, particularly in food chains, losses always occur and the lost energy tends to appear as heat energy generally unavailable for further use.

CHAPTER 2

Population and Food

It is possible to describe the basic principles and processes of ecology without reference to man, and much of the content of chapter 1 would remain unchanged if man had never been present on the Earth. However, man is present and in large numbers. He is unique among the Earth's living things in the respect that he has the capacity to think about his environment as though he was an independent, detached observer, while at the same time his activities have the effect of modifying natural cycles and changing the environment on a scale far greater than that of any other animal. The extent of his impact will depend on both the nature of these activities and the sheer number of people engaged in them.

Many writers contend that the present world population is dangerously large and, more important, that the present rates of growth are so great that only short time intervals are needed to reach any of the figures estimated, even by optimistic observers, for the maximum number that the Earth can support. Such contentions lead to campaigns to limit population growth and often contain proposals that cause offence because they involve intrusion into those areas which we regard as private and personal, as well as those which concern social, educational, and taxation policies. Some of the basic facts and principles of population studies can be established, and must be, if the discussion of the social and personal consequences of particular proposals is to be a well-informed discussion.

2.1 The Growth of World Population

The graph in box 2–1 (figure 5) shows clearly the dramatic increase in growth rate which began during the period 1650–1750. One way in which the growth rate can be studied is by considering the percentage increase per annum, and values for this quantity have been set out in box 2–1 (figure 6) for

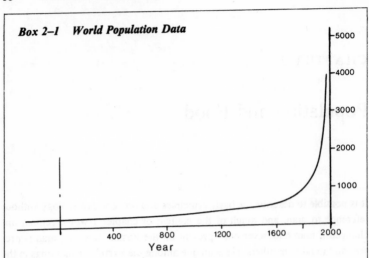

Figure 5   World Population (millions). (Values based on estimates by Bogue, 1969[1])

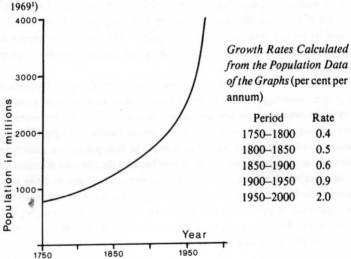

Growth Rates Calculated from the Population Data of the Graphs (per cent per annum)

Period	Rate
1750–1800	0.4
1800–1850	0.5
1850–1900	0.6
1900–1950	0.9
1950–2000	2.0

Figure 6   World Population 1750–2000 in Greater Detail. (Values based on United Nations 'median' estimates, 1968[2])

Doubling Time in Years Associated with Particular Growth Rates

Growth rate	0.1	0.5	1.0	2.0	5.0	10.0
Doubling time	700	140	70	35	14	7

Box 2–1   *World Population Data*

the period 1750–2000. Before 1750, on a global scale, the growth rate was probably never greater than 1 per cent per annum for any significant period of time. This kind of growth characterised by apparently small numbers in the early stages subsequently generates very large numbers very rapidly. In its simplest arithmetic form it is exemplified by compound interest rates where interest, added to the original capital, itself attracts interest in subsequent periods. With populations, extra children grow into child-bearing adults, and the growth has obvious similarities, though, of course, the net growth rate of a population is the difference between births and deaths; some capital goes out of the system. Growth at a fixed rate of compound interest is known as exponential growth and it is a common feature of biological, physical and financial systems. A legend which illustrates the distinctive characteristics of this kind of growth speaks of a courtier who presented a king with a chessboard and asked to be paid in rice grains. One grain of rice was to be placed on the first square, two on the second, four on the third, eight on the fourth and so on, doubling each time. The king agreed readily and started to pay out. The tenth square took 512 grains, the twenty-first square more than a million grains, and the fortieth square a million, million grains. The process could not be completed with all the rice in the world. The regular increase with each step is equal to the preceding value and could be described as 100 per cent increase per square. The arithmetic of exponential growth is shown in box 2–1 where the rates of increase are per year and are typical values for the growth in population or growth in demand for energy, fertilisers, waste tipping space and other features of life in the twentieth century. When we examine how the world population has grown even during that relatively short part of man's history that we call the Christian era we see that the baseline value of 250 million was doubled during the 1500 years that followed; the next doubling (500–1000 million) took only 310 years (1500–1810) and the next (1000–2000 million) only 120 years (1810–1930) and the next (2000–4000 million) has taken only 45 years (1930–75). Clearly the doubling time has been decreasing rapidly as the rate of growth has risen. At the time of writing the growth rate is slightly above 2 per cent per year giving rather less than 35 years for the doubling time. Growth at a growing, rather than a constant, growth rate is not strictly exponential and could be described as super-exponential.

Within these global totals there exist important differences between geographical regions. The growth in population in those parts of the world known as the more developed countries took place during and after their industrial revolutions; the far more numerous countries which are labelled less developed are experiencing their population surge before their industrial revolutions. Box 2–2 (tables and figure 7) shows more detail of the population distribution. It seems that during industrialisation birth rates in Europe were more or less constant and death rates fell, but after industrialisation both birth and death rates fell. So Europe has moved from its position as growth leader in world population tables from 1750–1850 to its position during the twentieth century where its growth rate is below the world average. By contrast the less-developed regions of the world have generally been characterised by slow growth rates right through the early part of the period under consideration, and until we reach the decade 1930–40 when we see a really dramatic decline in death rate as a result of the flow of medical

---

Box 2–2   *World Population: Geographical Distribution*

World Population Distribution (millions)

Year	1800	1850	1900	1950	1975	2000
More developed	160	341	562	836	(1100)	(1450)
Less developed	820	918	1088	1684	(2900)	(5040)
Total world	980	1260	1650	2520	4000	(6490)

Note 'More developed'—definition changes during period covered, but in second half of the twentieth century means Europe, U.S.S.R., North America, temperate South America, Japan, Australasia.

Detailed Distribution by Continents (millions)

Year	1750	1850	1950	2000
Asia	497	800	1385	3756
Africa	106	111	222	817
Europe (excluding U.S.S.R.)	125	208	392	568
U.S.S.R.	42	76	180	329
North America	2	26	166	333
South America	16	37	162	652
Oceania	2	2	13	35
Total world	790	1260	2520	6490

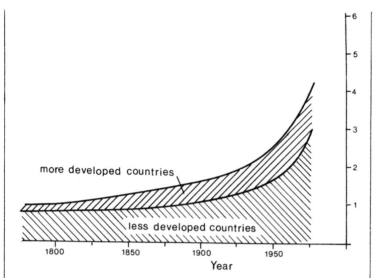

Figure 7   Graph of Distribution of World Population Between the More and the Less Developed Countries (1000 millions). (Data based on United Nations 'median' estimates, 1968, and on J. Durrand, 1971[3])

knowledge and techniques both in curative and preventive medicine, from the more developed to the less-developed countries. The collapse in death rate can be seen by considering examples such as Sri Lanka (Ceylon) where in 1935 the death rate was 34 per thousand; in 1945 it had dropped as a result of generally improved health and medical care to 22 per thousand. In 1945–7 a major assault on malaria (one of the principal causes of early death) reduced the death rate so rapidly that in 1948 it had fallen to 14 per thousand, by 1954 it was 10 per thousand, and, in 1971, 8 per thousand. This decline in death rate is most apparent in the saving of young lives but has not been accompanied by any similar decline in what are still high birth rates. So the population of the less-developed countries continues to grow. Graphs in box 2–3 (figure 8) show the contrast between a less and a more developed country (typically Sri Lanka and Sweden).

If we consider in greater detail the population of a less-developed country we find that one of the effects of declining death rate and constant birth rate is to produce a situation in which a large fraction of the total population is dependent, that is, too young or too old to contribute significantly by working to support themselves. From these considerations of age structure a further

Box 2–3   *The Changing Pattern of Birth and Death Rate*

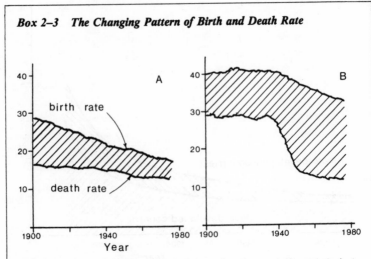

Figure 8   Pattern of Birth and Death Rates (per thousand of population). A, for typical developed country; B, for typical developing country

Advances in medicine have brought down death rates throughout the world. In a typical developed country (A) the fall has been gradual during this century; in a typical developing country (B) the fall has been dramatic and concentrated into the past 30–35 years. The rate of growth of the population is governed by the gap between birth and death rates. Birth rates have fallen in both types of country, but the gap between the rates is increasing in (B) and decreasing in (A).

important feature of population emerges; populations are subject to inertia and change only slowly from their established trends. The population of the world in 1990 is already determined because the child-bearing part of the 1990 population is already alive in 1975. During those fifteen years a developing country with a net growth of 2 per cent will increase its total population by one third. Changes in the birth rate, or death rate, which would make this forecast seriously in error suppose changes in the former which are most unlikely or in the latter on a disaster scale worse than the world has yet known.

Looking into the future as far as population figures are concerned is necessary, difficult, and at times politically explosive. The term generally used for such work is projection (rather than forecast or prediction) to emphasise that statements are made about the consequence of continued growth at the present, or some other, specified rate. They are concerned with the effects of carrying forward clearly stated patterns rather than unconditional assertions about future population sizes. The United Nations figures are offered in a variety of forms. They are not simple projections based on the current growth rate but take into account detailed data on specific age fertility and death rates both as they are at the time of the data collection and as they might be if the variations already observed in similar situations in the recent past take effect. Major disasters on the scale of thermonuclear war are not however included. These lead to four sets of figures: the high and low projections, a medium projection, and a *constant fertility, no migration* projection. The 1968 projections show the world population figures for A.D. 2000 as, low projection, 5449 million; medium, 6130 million; high, 6994 million; and the *constant fertility, no migration* projection yields the highest figure of 7522 million.[4] These figures were published in 1968, and the intermediate figure within the U.N. projections which most nearly agrees with the actual 1975 world population is that corresponding to the 'high' estimate. Projections made over the last thirty years have tended to err on the low side and the U.N. estimates for world population in A.D. 2000 have been steadily increased as the end of the century draws nearer.

2.2 Population Dynamics

Many of the historically important studies of population dynamics have been carried out on animals other than man and particularly on animals which have a much shorter life span. These studies have shown that populations of such animals are subject to controls which operate in combination although one controlling factor usually dominates at any given stage. The simplest model (which will need subsequent modification) supposes we have, in controlled laboratory conditions, a starting population of small animals and we carry out the experimental work in such a way that no regulating factor is allowed to operate. If we start with an all-adult population there would be a high initial growth rate followed by a period with no increase until the offspring reached the age when they, in their turn, were able to reproduce. There would be series of surges of population which would gradually smooth out until a more or less steady rate of growth was established. Starting the experiment with eggs leads to similar surges but starting with a mixed-age population would lead more quickly to a steady growth pattern. The graph of population size would be similar in shape to the graph of human population of the world as shown in box 2–1. The laboratory experiment, however, lacks reality, and we know that in the real world controlling factors are present and we can make good guesses as to the nature of these factors and so develop the experimental work that one particular interrelationship can be isolated because other factors are not allowed to regulate. Factors which have been explored in this way are food supply, living space, pollution, disease, predation and parasites.

From this evidence it is possible to establish a general picture of population control in plants and animals, which uses the idea of regulating factors which are *density dependent*, that is, as the population density increases these natural factors have an increasingly severe effect whilst when population falls their effect decreases. Density-dependent factors operate in such a way as to increase the death rate, decrease the birth rate or stimulate migration. We must then ask what natural regulating factors appear to operate as far as human populations are concerned. For man as a species we can conclude that he has the capacity to reduce the effects of the factors so far described. Predation on man is negligible since the development of long-range weapons and hunting skills; parasites are combatted by drugs; hunger is less likely to affect man with his varied diet and his skill in farming and storing foods and most killer diseases are under control. It could be tempting to infer that man as a result of his intellectual and technical superiority has the capacity to evade the natural regulating factors which in plants and animals work effectively to control population size. However, it is also true that, in many long-established traditional societies, social restraints operated which resulted in regulation of population size; these included tribal rites and taboos and inheritance patterns which led to late marriage and reduced fertility.

In discussing human population control we move into a controversial area. Control of population implies the choice of optimum and maximum populations against which present or anticipated populations may be compared. Such values are the result of judgements about the capacity of the agricultural system to support people and about such issues as the definition of subsistence, welfare and quality of life, and these are not merely matters of mental arithmetic. There is already general awareness that some millions of those alive today are suffering and will die as a result of an inadequate supply of food. Any optimum population figure which assumes that the distribution of the world's food output is based on need rather than ability to buy must be treated as unrealistically optimistic and any other optimistic assumptions about social redistribution must be balanced against a situation characterised by a rate of growth which doubles the world population in only 35 years and so gives little time to manoeuvre. There is general agreement that the size of the world's population and particularly the rate of growth is too large for the health of mankind but there is marked disagreement on the implications and on the remedies. Broadly, two schools of thought emerge. One sees population control as essential for survival and speaks of the threat which the explosive growth of population has brought. Because the time available is so short drastic measures are called for and population control, even with coercive elements if necessary, is inevitable and urgent. This position is taken by a large number of scientists and political observers in the developed countries and is seen as applying particularly to the people of the developing countries (though most would agree on a less severe and urgent need in their own countries). It emphasises the thesis that growth of population on a finite Earth with a limited support capacity can only lead to disaster; it sees that disaster as a real threat within the next 100 years during which period unhindered growth at present rates would lead to an eightfold increase in population size.

The other extreme view is that a frontal attack on population is both unjust and ineffective. Population growth will only be slowed when development leads to major changes in the distribution of wealth—when we look after the people the population will look after itself. Population control is only effective where it has the active support of the people and in many of the countries with high growth rates people have large numbers of children because they want large families, and they want large families because they see them as necessary for survival. This view is held by many of the radical thinkers from the West and from the developing countries, who see the population control remedy as one which arises from a desire to retain the present world distribution of wealth. Development policies and population policies must go hand in hand but the dominant issue is seen as development. Development may seem a luxury to those preoccupied with the struggle to achieve bare survival. In the next section of this chapter we shall consider the demands that man makes on the natural world around him, first in terms of mere survival, and then, in chapter 3, of the demands implied by the use of the term development as they appear in the developed countries today.

2.3 The Basic Needs for Survival

At the most fundamental biological level man needs food and water, shelter and warmth for survival. His food provides him with the energy he needs for both growth and activity, and for the replacement of body tissue. Considering only his energy needs we find that, at the lowest rate of energy consumption, while resting, a normal adult needs 65–85 W (about the same power demand as a reading lamp); if he undertakes any activity the demand rises sharply. Averaged over a day the need is for a rate of supply of energy of about

125 W as an ordinary minimum for moderate activity and this clearly must match his energy input from carbohydrates and fats if he is to sustain this level of activity. In addition his foods provide other essentials for life, one of which, protein, can in fact be used to make up any energy deficit that the fuel foods already mentioned are unable to meet. Protein is an essential for life as it provides the structural materials of the human body, and serves as a biological catalyst as well as acting as the oxygen transport material; the bulk of the cells of our body are protein. All of the enormous range of proteins are built up from the twenty or so amino acids in different combinations; eight of them are essential to adult life (one further one is necessary during childhood) but from these nine the body is able to build up the others. Dairy products contain all nine amino acids in about the correct proportion to meet our nutritional needs. Meat and fish are also good sources, while soya bean and some nuts provide all the proteins but not in the proportions needed. Cereals and some vegetables and fruits provide several of the amino acids though no one of these sources can meet all man's needs. Vitamins are essential, in very small quantities compared with carbohydrates or proteins, and have a controlling role in the body processes. About thirteen different vitamins have been identified. It is also possible to identify about seventeen minerals which also seem to be essential for life; these include calcium and phosphorus, salt, iodine, iron, sulphur, potassium, zinc, magnesium and manganese.

The demand for food can only increase in the foreseeable future and there are relatively few devices available which can lead to an increase in supply. The *per capita* food production considered on a global basis has risen only very slightly over the last thirty years, the years that have probably seen the most intensive high level activity on food production, and over the last ten years there has been no improvement in this figure. This means that, while the proportion of the world's population which suffers from malnutrition remains about the same, the actual number of people suffering is growing as rapidly as the population as a whole, and this represents an enormous increase in human misery.

One major improvement in the situation would result from a more rational and humane distribution of the food that is available. The past 30 years have seen many instances of starvation occurring in some parts of the world while excess foods are destroyed or ploughed back into the soil in others. If, however, we confine our attention to increasing food production we have the possibility of increasing the amount of land under cultivation, increasing yields, reducing losses due to pests and winning more food from the sea.

At present about 15 million km² are under cultivation and this is estimated to be just under half of the potentially farmable land on the Earth. A further

29 million km² are used as grazing land. There is considerable doubt about the extent to which these areas could be increased without enormous cost even though we have called this remainder potentially farmable. The tropical rain forest looks rich and fertile but, because it is mature and has a very large mass of living material, a larger fraction of the nutrients in the system are in the plants rather than the soil. When the trees are removed on any large scale, exposure to sun and wind, deprivation of decaying plant material and changes in moisture content much more extreme than any experienced under the canopy of trees, result in a process known as laterisation. The heavy rains leach out nutrients and within a few years the soil is brick-hard and unproductive; this process can be seen in most tropical areas and is still taking place. Forests cover 45 million km² of the land surface of the Earth and with care the food production from this area could be increased while it is still forest by the use of tree crops or shade-loving crops. It is significant that the greater part of the ice-free land area which would bear investigation as far as increasing the food producing area is concerned is in developing countries which lack capital and technical expertise. The facts appear to be that given our present agricultural technology and economic system we are using all the land we can and that in fact we are losing land faster than we are bringing marginal lands into production. The deserts are certainly advancing each year. The Sahara extended its area by about 40 000 km² during the severe years of drought in the 1969–75 period. Heavy use of good land, overgrazing of marginal and pasture lands still leads to loss of valuable topsoil when weather conditions are severe.

The next two possibilities were increasing yields and reducing losses due to pests; before we can explore these we need to look at the basic processes of agriculture and its recent technology. The plant itself is a growing organism and each has its own particular demands that must be met if it is to grow in a healthy way and to fix solar energy. These nutrients must be available in the soil in which the plant grows, in sufficient quantity at the particular times in the growing cycle when they are needed by the plant. This again emphasises the cyclic nature of the process with the flow of nutrients through their cycles controlling the output of particular plants, unless man deliberately short-circuits the system by mining a particular nutrient at another place, transporting it and feeding the soil in the fields he is cultivating to remedy some deficiency. Historically, time was gained to allow the natural cycles to replenish the soil by rotation of crops with different nutrient needs and by leaving land to lie fallow; the increased pressure on farmers now means that these techniques are less attractive in the shorter term, as are techniques which involve enriching the soil with some of the decomposing plant material which was not taken for other purposes. This would give back to the soil some of the material originally taken from it as well as providing material which would keep the soil well broken up and improve its water-holding and aerating qualities.

The water needs of plants are often crucial. Plants are prodigious users of water. Typical figures show that the use of water to provide the hydrogen atoms that are bonded into the carbohydrate molecules of the plant accounts for a tiny part of the water demands of the plant during the season. For each kilogram of fresh plant material about 100 kg of water will have been drawn from the soil during the growing season. At harvest 75 per cent of the fresh weight may well be water in transit and only about 15 per cent of the fresh weight is fixed water. This means that of the total water demand only 0.15 per cent remains as part of the plant material after harvest. The water in transit has been drawn from the soil, through the roots, passes through the plant and is evaporated from the leaf surface.

One other fundamental need of plants is nitrogen. The atmosphere is almost 80 per cent nitrogen and at first sight this need would seem unlikely to be crucial with such an abundance available. However, nitrogen in the atmosphere is not available to almost all plants and animals in its gaseous form. It must first be fixed by being combined chemically with other elements before it can be used. A very much simplified picture of this process shows the gaseous nitrogen molecule splitting into the two separate nitrogen atoms and then being combined with hydrogen (or oxygen) through the activity of micro-organisms in the soil. The nitrogen compounds are then in a form which plants can assimilate. When the plant dies the nitrogen compounds may remain in the soil available as a nutrient for other plants or return to the atmosphere as gaseous pure nitrogen. Over the past 70 years industrial processes have been developed which enable us to fix atmospheric nitrogen and produce nitrogenous fertilisers. The natural fixation of nitrogen in the soil occurs most significantly in the nodules on the roots of the class of plants known as legumes (for example, beans and peas).

Some of the world's apparently potential fertile land has turned out to be unusable because it has been deficient in some mineral element (for example, copper or zinc) needed in such small quantities that it is difficult to detect and measure. These values are of the order of 0.2 g/m^2 but in the absence of these trace elements the otherwise fertile land can become a desert in which little or nothing can be grown.

2.4   The Techniques of Agriculture

In considering the techniques of agriculture we shall centre our attention on the developments of the past 50 years, the period which has been most influenced by advances in science and technology generally and the period during which the communication of techniques has been more rapid than in any earlier stage of its development. For convenience these aspects of agricultural technology are treated under five headings: breeding, irrigation, fertilisation, mechanisation and pest control.

(1) *Breeding*. Although selective breeding of plant and animal species has been an important agricultural technique through thousands of years, the development of the science of genetics has accelerated the process. It is now possible to give a specification for a new plant or domestic animal and to stand a fair chance of breeding it within a few generations providing the specific characteristics are already exhibited by common strains. Thus the farmer's search could be for a variety of cereal which is higher in yield, but also has a shorter, stiffer stalk, is more resistant to drought, more tolerant of temperature extremes, and less easily damaged by disease. In the 1930s a hybrid corn was developed in the United States which had several attractive features for farmers, and which produced greatly improved yields, because it was more robust and had a more extensive and vigorous root system—to name but two distinctive characteristics. In addition its season was shorter which allowed it to be used further north. With its use a 60 per cent increase in yield resulted during the period 1935–45. With domesticated animals there has been a similar increase in yield following selective breeding, particularly in terms of egg or milk yields which have been increased by ten to fifteen times over the yields of their original ancestors. In addition man has transported and crossed strains from different parts of the world, so modifying the relative abundance of the varieties and their geographical distribution.

All this has not been done without cost. The hybrid corn yielded well in terms of starches, sugars and fats but the increased yield was accompanied by a fall in protein content to a half of the value normally expected from corn before the high yield crops were introduced, and well below corns which were generally considered as low yield varieties. In addition, the hybrid extracted more from the soil which meant that a much heavier input of artificial fertilisers was needed, or alternatively, the soil needed resting for longer fallow periods. There seems to be a fundamental guide that the higher the yield, the lower the nitrogen content of the crop (and this generally means less protein value).

(2) *Irrigation*. The dependence of growing plants on a plentiful supply of

water was discussed in an earlier section of this chapter. From earliest agriculture man has found ways of supplementing the natural supply of water from rainfall, by utilising flooding patterns of rivers and subsequently by building an artificial network of small channels across his cultivated areas and diverting natural flows into them. Basically they diverted water to the soil and hence the crop, to allow supply through that part of the dry season when the plants needed it for growth, and also to extend the area on which crops could be raised. Water thus diverted goes back into the cycle through evaporation from the soil, transpiration through the plant itself (the real objective of the exercise) and, if the geological pattern permits it, by soaking downwards into aquifers. In more recent times we have seen the beginning of a catastrophe, in Pakistan, where, in 1961, it was stated that an agricultural land loss of 250 km² per year was due entirely to a side effect of small-scale irrigation.[5] It seems that the diversion of water from the Indus, over a period of more than a century, on to the fertile plains of West Pakistan, had raised the water table (as a result of percolation through the soil and collection in underground reservoirs) until it was less than a metre below the surface. As a consequence roots were waterlogged but, in addition, there was an accumulation of salts in the soil. When slightly salty water is evaporated the salts remain and over a period salty deposits accumulate so that the soil is eventually useless for agriculture. In this particular case the solution eventually found involved drilling large numbers of small wells and pumping water from the underground reservoirs for irrigation purposes, rather than solely from the river. This lowered the water table and washed the salts deeper, below the actual topsoil. This land is steadily being reclaimed; the extent to which the Sahara desert was the result of salination and waterlogging is not clear, but it could well have been a major cause.

In contrast the past 30 years have seen a large number of major schemes to dam rivers on a large scale, providing irrigation and power through hydroelectric schemes. In the developing countries these are usually fairly straightforward dams with their reservoirs, but on a much larger scale than previously known, and have necessitated a massive inflow of foreign capital and expertise. These large areas of water, in tropical countries particularly, have brought local changes in climate but, in addition, the flow of silt (part of the natural transportation process which normally takes place in rivers) does not stop, and the reservoirs silt up, which gives them only a limited lifetime. The other feature which has become apparent is that the large surface of still water has provided an ideal environment for aquatic weeds. The dam on the White Nile near Khartoum in the Sudan enclosed clear waters in 1958, but within months of its first appearance, huge mats of water hyacinth began to cover the surface of the enclosed water. The weed mats were sufficiently dense and extensive to block the use of the lakes for transporting and fishing; the fishing grounds were seriously damaged because so little sunlight penetrated and the oxygen content of the water fell. The actual weed material clogged hydroelectric plants and blocked irrigation channels but, above all, the loss of water through transpiration by the weeds represented a very serious depletion of the reservoir capacity. A mat of water hyacinth transpires water at a rate between three and eight times the corresponding rate of evaporation from open water. It is also worthy of note that major irrigation schemes have spread not only water but also disease. Bilharzia, sometimes known as snail fever (because the parasite that carries this seriously debilitating disease is carried by small freshwater snails and burrows under the flesh of people who work in the irrigated fields), is an example of this process. Bilharzia now seriously afflicts 250 million people and is spreading rapidly.

(3) *Fertilisers.* In a wild situation the nutrients which are taken out of the soil by plants are returned to the same soil when the plant dies and decays. The cycle is a small-scale, rapid one in which soil quality is maintained. In agriculture the plant material is taken away from the soil which produced it at harvest, and with it the essential nutrients, so that if no effort was made to replace them the soil would deteriorate until it could produce no more vegetation. If a good supply of decaying plant material and animal waste is returned to the soil it can be maintained in a fertile condition. Thus, one stage removed from the natural wild condition, soil can be enriched by the deliberate return of material taken from it, but skill and understanding is needed to maintain a good balance. The natural cycle is managed and only slightly modified. Leibig, at the beginning of the nineteenth century, carried out a careful analysis of the chemical requirements of fertile soils and showed the specific needs for such elements as nitrogen, potassium, phosphorus and sulphur. As population pressure increased, new lands were opened up to agriculture and higher yields demanded from hard-worked soils. The use of industrially produced compounds which contained these elements, the inorganic chemical fertilisers, grew. The use of these fertilisers on a large scale began after the Second World War and has grown extremely rapidly since, so that at present over 70 million tonnes are produced each year. This total is made up of three main types: nitrogen, in the form of nitrates, phosphorus, in the form of phosphates and potassium, as potash. The world consumption of nitrates roughly equals in quantity the sum of the other two and is still rising steeply. There is no doubt that where fertilisers are skilfully used large increases in

yield result, and where there is sufficient rainfall and plant varieties are bred for this purpose, triple yields are possible. This is the key to understanding the achievements of the so-called Green Revolution. The demand for fertilisers is, however, enormous, and the cost of purchasing them beyond the means of those who need the improved yield most. The long-term agricultural consequences of the Green Revolution (not to mention the social consequences arising, for example, from the widening of the gap between the rich large-scale farmer and the poor small farmer) are not entirely beneficial. We have already mentioned the cost in terms of protein content associated with high-yield crops; there seems to be a cost in terms of resistance to disease or pests which seems to indicate that we cannot, as yet, be confident about our plant breeding technology and its capacity to meet all the specification requirements in any one plant type. Nevertheless there is certainly real gain already especially in the face of existing desperate famine conditions.

There is the important side effect of replenishing the essential elements in the soil with chemical fertilisers, which must be studied. When the replenishing of the soil is carried out by returning the nutrients as constituents of plant and animal tissue, they are fixed, chemically compounded in the large molecules of the material. This means that very little is washed away by surface water runoff. In addition, the natural manure has large fibrous and bony particles which help to give the soil a good texture, allowing water and air to flow to the roots and allowing the roots to penetrate the more open structure. The function of this humus in soil is vital to its capacity to go on producing harvests. If, however, the right nutrients are supplied to the soil in the form of chemical fertilisers the chemical needs may be met but there may be a lack of humus to give the soil its necessary texture, and there is the increased loss of nitrates and phosphates which are washed from the soil during heavy rain. These fertilisers find their way into streams and lakes and enrich them as environments for plant growth. The consequence can be an explosive growth of algae which, in turn dies and decays, making increased demands on the oxygen in the water of the lake, killing off the fish and other water animals. This accelerated ageing is known as eutrophication; many lakes and inland waters have suffered badly already and Lake Erie has been the classic example of this. In fact the nutrient supply to this lake from agricultural runoff was not the only, nor even the major cause of this eutrophication; the major contribution came from industrial and domestic sewage.

(4) *Mechanisation.* The developed countries have embraced a machine technology in agriculture to the extent that, over the past 30 years, there has

been a reduction to almost one-third in the number of farms, and a fall in the number of people working on farms from about 25 per cent of the population, to about 6 per cent, without any significant decrease in the total area farmed. This has been made possible because labour and capital are employed in producing equipment and machinery to extend the handling capacity of the farm worker, and because the concentrated energy services available in the fossil fuels, have been drawn into farming. As a result of this trend agriculture has taken more of the pattern of factory-based industry, with greater regularity and uniformity; fields are larger, furrows straighter, livestock units are larger and have more mechanised handling so that biological systems are being strained into a machine-age form. It we attempt an energy audit for a typical farming production unit we find that 50 per cent of the energy input is represented by fertilisers while fuels (mostly oil products) account for 30 per cent and human labour input is less than 1 per cent. Clearly there is a major fossil–fuel subsidy in food production. In some highly specialised sectors of agriculture the fossil–fuel input exceeds, in energy content, the value of the foodstuff produced, when we add in the fossil–fuel cost of the production and distribution of machinery and fertilisers. For a variety of reasons a developing country, without its own oil, cannot consider such a possible path of development. The advantages of mechanisation are obvious; the costs are not so obvious but are nevertheless very real. The increased use of machinery compacts the soil, especially if it is low in humus, and spoils its drainage properties in a way draught animals never did, and it tends to move agriculture away from mixed farming where the manure from animals is the organic fertiliser for the plants.

(5) *Pesticides.* The last technology on our list is pest control. The destructive activities of insect pests is a major loss item on modern food production and Georg Borgstrom, using United States Department of Agriculture figures, estimated that, in 1969 the field loss of total production is about 10 per cent and the storage loss is about the same size.[6] This agrees with estimates made 20 years earlier and seems to indicate that the enormous increase in the use of pesticides has not improved the situation. The overall impression gained from case studies is that larger and more powerful doses of insecticide need to be used in successive years as insects develop resistance and as their natural predators are killed off by the chemicals used. The larger fields of more uniform crops mean easier harvests for the insects and the impression is gained that very few systematic large-scale applications of chemical pesticides have been really successful. Pesticides have other effects than killing the specific pest. The bulk of those used at present are synthetic

hydrocarbons which have four important properties. They are toxic to a wide range of organisms including vertebrate animals, they are chemically stable and do not easily break down to harmless components, they are very easily transported in the atmosphere and they tend to accumulate in the fatty tissues of organisms. The evidence of the experience gained with D.D.T. (probably the best known of this group) since it was first used on a large scale early in the 1940s, is that it persists and has spread over the whole globe. It has had a major role in the control of insect-borne diseases and it has kept control of pests in a wide range of agricultural applications but it has also accumulated in the tissues of most animals, including man, to a level which some contend already makes it a serious health risk and one which, because it is cumulative, is very difficult to assess. The chemical effects of these pesticides on, for example, insect-eating birds, and on predatory birds shows itself in increasing concentration in the tissues as we go further up the food chain. However, even more serious is the disruption of natural cycles of which these effects are only symptoms.

There is an alternative approach. Insects which we call pests because they compete with us for food supplies, or carry diseases, are themselves subject to predation. Many are threats to agricultural production because they have escaped from the control of natural predators. A fuller understanding of their place in the natural web of interaction can lead to the identification of predators that can be used to exercise biological control. In other situations we find that the insect species is particularly vulnerable at some early stage in its life cycle and we can effect biological control over its foods and environment to reduce its numbers at this stage or alternatively we may find a suitable parasite which controls the insect population effectively. There is now more than sufficient evidence that biocontrol is safer and more effective than chemical control and shows extremely good cost effectiveness, but it does take time to research and to have effect.

As an additional device to feed a hungry world to add to those which have been discussed already, the idea of increasing food yields from the sea seems attractive. The vast oceans contribute a small (but not negligible) amount of our energy foods but make a contribution to our protein supplies of about 20 per cent of the total animal protein. It is, however, more accurate to describe man's activities here as being more like hunting than either herding or farming and to observe that this is far from primitive hunting since it is served by powerful modern technologies. The seas can be over-fished and because these technologies are so powerful and effective we are in danger of exploiting a short-term opportunity in a way which cannot be sustained. The whaling industry provides an example of this: three of the larger varieties of whale have been hunted to the point of extinction, while a fourth is similarly threatened. The whaling fleets catch more smaller and younger whales each year but the total tonnage caught is falling. Some countries are now harvesting the small shrimps which are the main food of the whales, which seems rather like killing the goose that lays the golden eggs. The facts about the biology of the seas seem to lead to the conclusion that the deep ocean is effectively a desert and that plant production in shallow waters is of little direct use for man. So man eats at higher levels in the food chain. Photosynthesis only occurs in the top metre or so and productivity is low because nutrients are generally in short supply; only in coastal waters is productivity at a reasonable level and we take our fish from these levels now. Pollution is a major problem in most of them already and is increasing. It does seem that while through fish farming and carefully controlled harvesting we might increase yields on a sustainable basis the sea will not make a major new contribution to the world's food production.

Changes in our eating habits could be important. At present we import protein from developing countries to feed to animals in the developed world to provide meat, and we are thus able to sustain a diet which gives us food from high levels in the food chain. As food shortages develop this will become a less tolerable privilege. The soya bean is principally used as a protein food for cattle but does have considerable merit as a protein source for man; at present, in the developed countries, research efforts are directed towards making it look and taste like meat in order to overcome the prejudice against it as an inferior substitute. Feeding habits are notoriously difficult to change and this has been particularly evident where groups of people are near to starvation; in fact there is far more tolerance of novel foods among people who are well fed and accustomed to a varied diet.

Summary

In this chapter we have considered the dramatic increase in the world's population which has occurred over the past 200 years, and the changes in birth and death rates, which at differing times in different parts of the world, have produced it. The natural checks on population growth which operate for animals and plants have been discussed together with the evidence that man has been able to delay or avoid them. In a book of this size it is not possible to discuss the techniques of birth control and contraception, nor to consider the effects of abortion (both spontaneous and induced) which is the largest single cause of maternal death; it is estimated that between a third and half of all pregnancies end in abortion. These subjects are more fully treated in some of the books listed in the Further Reading section (page 158). The attitudes to population control and likely growth of populations have led on to questions about the pressure which this will place on the capacity of the Earth to meet even survival needs. We have then considered in detail the particular survival need for food both in terms of the basic biology and the techniques of agriculture with special reference to plant genetics, irrigation, fertilisers, mechanisation and pesticides. In a world in which millions die every year as a result of an inadequate diet, improvements in production and distribution are urgently needed; it is hard to avoid the conclusion that large-scale famine and malnutrition are more likely to increase over the next 25 years.

CHAPTER 3

Energy, Materials and Pollution

Over the past 150–200 years there has been an enormous increase in both the quantity and the complexity of man's demands for materials and energy, taking them well beyond mere survival level. These increases have come to be seen as part of the process of development as we have come to know it over this period of history, and seem to characterise the developmental planning of both more, and less developed countries. We shall consider these demands in this chapter. In chapter 1 the use of the term 'consumer' was challenged and the idea of materials moving in cycles in which their interaction with man was seen as only a part of a larger process, was used instead. These cycles were seen to interact with one another in a complex but relatively stable manner and the energy which made possible both the chemical recombinations and the actual transportation was seen to flow through the ecological system. We shall develop this picture further by describing the three-stage process by which man modifies the cycles by his activities as an *extraction*, *manufacture and use*, and *disposal* process.

By *extraction* we mean the withdrawal of a material from the cycle at a particular point in order to make it available for processing, or the harnessing of energy as it flows through the system. By *manufacture* we mean the transportation, and the chemical or physical rearrangement and combination with other materials, and the distribution of the product to the consumer. The consumer then uses the material—the most obvious part of the process—and at some stage decides that the article is no longer useful to him and decides to dispose of it. This decision may be made before the item is worn out and may have been made with a desire to replace it by a more up-to-date and fashionable product. For whatever reason it eventually becomes waste. *Disposal*, the final stage in the process, generally results in the reintroduction into a natural cycle of a composite article quite different from anything found in nature. Looked at in this way we can see that the extraction–manufacture and use–disposal process has resulted in a new path within the general pattern of a natural cycle of a particular material which has taken it through a different series of chemical combinations, at a rate different from the natural rate of the cycle, and disposed of it in a different chemical form, a different concentration and in a different geographical location. The resilience and adaptability of the ecosystem is obviously a matter of some importance. The direct human use of foods, because we are dealing with a natural living system, is intrinsically less disturbing than say, the use of iron in a manufacturing process. Certainly this can be true when a farmer in a temperate climate grows his own food, eats it and disposes of the waste by returning it to the soil; the new path is short and has no particular dissimilarity in terms of rate of flow or of concentration as long as his land is not over used. In considering resources we shall consider *renewable* resources by using this term for resources which can be managed, that is, directed with regard to concentration in space and with regard to the rate of extraction and natural replacement. Few materials or resources fall into this category except solar energy, food crops, and water. Mineral resources cannot be included since their cycle is a geological process and must be seen as providing replacement materials only over millions of years; once the metallic-ore concentrations have been taken into the manufacture and use–disposal process (and the disposal is a comparatively rapid dispersal to a very low concentration) it will not recycle again for so long that for all practical purposes it must be classed as non-renewable. It is particularly significant that the newer demands associated with development have only been met where there has been an increased skill in harnessing energy. Normally, solar-energy income and geothermal-energy capital combine to drive the natural cycles of materials, but when these cycles are modified in some way which involves a change in the path followed, or the rate of passage through the cycle, a further input of energy is required. Metals move very slowly through their natural cycles and are relatively rarely available in high concentrations. To use them in industrial processes we must expend energy in searching for concentrations of the metallic ore, mining, refining, transportation of the metal, in manufacture, assembly and distribution of the product, in dismantling and disposal as waste after use. All these processes demand an energy source which is available in greater concentration than solar energy and which is easily controllable. Even the use of human muscular energy and draught animals in lead mining in the sixteenth century represented a short cut across a natural cyclic pathway and an increase in speed through that part of the cycle by a factor of several million. Until the industrial revolution the scale of such operations was generally small but the growth in demand for metals has been so rapid that it has been estimated that production since the beginning of the Second World War has exceeded all the production in the whole history of the world up to that time. Since energy in the form of concentrated fuels has made this rapid increase possible we shall consider first the energy resources available and the techniques of harnessing them.

3.1 Harnessing Energy

There is an important distinction to be made between energy and fuels. All fuels are sources of energy but not all sources of energy are fuels: fuels are materials which can be made to yield energy because the electron structure of the atoms (or the structure of the nucleus, in the case of the nuclear fuels) can be rearranged in suitable processes in such a way that some of the energy originally needed to maintain the structure can be released. Oil is an example of a chemical fuel in which the rearrangement of electrons during burning releases large quantities of energy. Fuels constitute the major sources of energy for the world at present and fall into three groups. The first comprises the foodstuffs (energy sources for man and other animals), animal wastes (typically cow dung which is an important domestic fuel in Asia) and wood. This group derives its energy content from photosynthesis in the very recent past. The second group comprises the fossil fuels, coal and oil, and the associated shales, tars and natural gas, which again derived their energy from solar energy by photosynthesis in the distant past, but which have also been concentrated by long-term geological processes which result in a higher energy content per unit mass. The third group is the nuclear fuels which derived their energy from much more violent processes which occurred before or during the formation of the Earth and which have an enormously greater energy yield. The energy from these fuels can, however, only be extracted by processes which are far more complex and dangerous than burning. Only the first group has the distinctive quality that the rate of production and the rate of usage are comparable and so only this group can be described as renewable. It is estimated that oil takes some 250 million years to form but at present world rates of usage about 10 tonnes are used every second; this difference in rates alone means that the fossil fuels are effectively non-renewable energy sources (the estimate for coal yields an even greater disparity).

The past 150 years have seen a change in the pattern of dependence as far as fuels are concerned. Until the seventeenth century, man utilised muscle power (his own and that of draught animals) and burnt animal waste and wood to meet his energy needs. By 1850 coal was established as an important world fuel, though its contribution was a little more than a quarter of the total, with wood remaining the principal source providing just under a half. By 1900 coal and wood had exchanged places on the fuel league table: coal use was increasing, but oil was not yet a significant contributor. Fifty years later oil had become as important as coal and its use was growing more rapidly. In the 1960s nuclear fuels began to contribute and their use is also growing rapidly. This changing pattern does show clearly that the world's energy needs are now being met almost entirely from unrenewable sources and raises the question of the length of time for which a world demand for energy growing at a rate of about 5 per cent per annum (doubling in 14 years) can be met from a fixed stock. The first simple answer is that it cannot be more than a few human lifespans for the fossil fuels unless there is a further major change in the pattern of fuel usage. To give more precise answers for each fuel source we need to specify a particular demand pattern and to have a reliable estimate of the actual quantities which might be available: these are not unrelated factors, since, for example, the extent of reserves of oil will depend on the importance we place on oil supplies and hence the price we are prepared to pay to find and bring to the surface oil in deposits which are remote, inaccessible and present technical difficulties. If we use the criterion of oil actually produced per metre of exploratory drilling we find that this has fallen over the past 50 years. On this basis alone it would seem that oil is getting steadily harder to find and the most optimistic figures for crude oil production and ultimate reserves show that we have already used about a quarter of all the recoverable oil there is likely to be available to us and that oil production is likely to reach a peak in about 1990 and fail off to a negligible value by about 2080. On this analysis the middle 80 per cent of all the oil produced will have been produced during the 70 years from 1955 to 2025. Viewed as a historical event on the whole timescale of man's activities this is a very brief episode. Applying the same analysis to coal, we have larger quantities and longer times but even here the middle 80 per cent of all the coal produced will probably fall within a 300-year period. This analysis can be studied in greater detail in M. King Hubbert's articles in *Resources and Man* (1969) and in the *Scientific American* (September 1970 issue). These figures are an indication of the length of time that these two fossil fuels can be expected to serve as major world energy sources. Fossil fuels have however been the driving force for a particular type of technological approach and development which has emerged relatively quickly and is very strongly established. They have immense advantages over wood in terms of energy per unit mass and over nuclear fuels in terms of simplicity and safety. Because they are so concentrated as energy sources they are dominant as fuels for transportation and are generally much more convenient in small-scale applications. But oil, for example, is not only a fuel; it is the raw material for fertilisers and pesticides, for the plastics industry and for many synthetics, construction materials and medical products. Its increasing importance as a material at a time when its lifetime as a world resource is obviously short raises considerable doubts about the wisdom of burning it and so destroying its unique chemical structure.

With the decline of fossil-fuel energy sources within the foreseeable future the liberation of large quantities of energy from relatively small masses of nuclear fuels has an obvious attraction. One gram of uranium 235 has the energy equivalent of 2.7 tonnes of coal so that an electricity generating station working at an efficiency of about 33 per cent and rated at 1000 MW would only consume 3 kg per day. The process of extracting the energy under controlled conditions is a large-scale, complex and comparatively dangerous one so that this is unlikely to be a fuel for individual transport or domestic use and its most obvious application is for large-scale electrical generation. Since 1956 electrical power has been supplied by this technique and nuclear power stations currently supply about one ninth of Britain's electricity and rather more than 1 per cent of the total electricity generated in the world. Vast sums of money have been spent on research into the use of nuclear fuels and the developed countries, generally, have expansion programmes for building new power stations and even developing countries are building or planning their own. The question of the lifetimes of nuclear fuels is obviously important. The simplest nuclear process is burning the uranium 235. This is a comparatively rare isotope of uranium and occurs as only one part in 141 of natural uranium so that the metal ore must be mined, and then subjected to refining and enriching processes before it can be used. Exactly how much is available on a world scale is difficult to estimate because, as with most materials, as the cost rises, so less accessible deposits become more attractive. Generally experts agree that the nuclear power programmes already published by the developed countries would call for an enormous expansion of the current rate of discovery at a level which is probably extremely unlikely to be realised. This view has strengthened the argument for a more sophisticated process in which a blanket of a uranium isotope which is not a natural fuel in a nuclear reactor (uranium 238), is placed around the reactor core and becomes converted into plutonium 239 which is a fuel comparable in energy yield to uranium 235. This process results in more fuel being produced than is consumed in the core, thus earning it the description of breeder reactor. The most optimistic estimate of the commercial production date for electricity generation by a breeder reactor is

1980 and the experience gained so far of nuclear power generation is that very rarely have these large-scale programmes fulfilled expectations. There are, however, major aspects of the use of nuclear fuels which should indicate a need for cautious development and these relate to safety. Nuclear reactors become unserviceable after about 30 years and then, because they are highly radioactive, they can neither be dismantled nor shifted but must remain where they are. The wastes which accumulate all the time the reactor is working are the most dangerous materials in man's environment and decay only over periods of time which must be measured in thousands of years. A single typical nuclear power plant of 1000 MW capacity produces a deceptively small quantity of waste in terms of bulk but its level of radioactivity is extremely high. The pollution aspect of this question will be considered in more detail in section 3.3.1 of this chapter but sufficient has been said already to show why the American nuclear engineer Alvin Weinberg describes nuclear power as a 'Faustian bargain' in which man is able to use a prodigious source of energy but only at the cost of eternal vigilance.

Fuels can only be usefully used as energy sources where we have the technology for liberating the chemical, or nuclear, bond energy in the place we choose and at a suitable rate. In chapter 1, comment was made on the efficiency of such processes, and on energy conversions generally. In most applications of fuels efficiencies are low; only in the case of conversion to thermal energy for space heating can we regularly expect efficiencies greater than 50 per cent. In the developed countries our use of fuels leads to a distribution of energy usage which, taken in increasing order, falls into four major categories—waste and losses (15–20 per cent), domestic and commercial (20–25 per cent), transportation (20–25 per cent) and industrial (35–40 per cent). To achieve these end products we first burn the fuel, then convert the greater part of the thermal energy into mechanical energy and just under half of the energy finally undergoes conversion to electrical energy. Most of the losses occur in the first two stages. Electrical energy is attractive because it is clean, convenient and easily distributed; it cannot be stored, except in chargeable cells (accumulators) as chemical energy, and the overall efficiency of the generating process is low (about 30–35 per cent). In the case of transportation and industrial applications (to a lesser extent) we use mechanical energy directly. We thus have an analogy with the food chain and it is not surprising that when we generate electricity from a fossil-fuel source and then use that electrical power for space heating or even for driving a vehicle we are going to do so at a high cost, in energy terms. Whether this cost is apparent in the price charged per useful unit of space heat energy may depend on subsidies or taxes adminstered to direct the pattern of consumption.

There is no doubt about the energy accounting, however; the direct process of burning fuel for space heating is twice as efficient as the process which uses electrical energy as an intermediary. Just as there are advantages in feeding at the lowest possible level in a food chain so there are advantages in using energy with as few intermediate conversions as possible.

What sources of energy, other than fuels, are available to man? The first group of resources derives from solar energy. The nuclear fuel in this case is being burnt in the Sun's reactor and we receive the radiant energy largely as heat and light. Clearly this is excellent for heating and lighting and human existence depends on it. It also provides energy for photosynthesis and thus for plant (and fuel) production. It does more than this, however. The uneven heating of the atmosphere causes convection currents which lead in turn to the movement of air masses across the Earth's surface: this energy has been utilised throughout recent history through windmills and sails. The heating of the Earth's surface also leads to evaporation which combined with the convection currents drives the water cycle and leads to large masses of water flowing back to the sea in rivers and streams. This again is already utilised in watermills and turbines. In recent years it has been possible to convert sunlight into electricity directly through solar cells (though efficiencies are, at present, below 10 per cent). This group of solar energy sources has important characteristics; the basic source is effectively unlimited, plentiful and has no adverse environmental effects. Taking the United States alone, the consumption of electrical energy in 1970 was around 2 million million kW h (this was about 10 per cent of the total energy consumption for the United States for that year). Sunlight falling on about 1000 km² of American desert land would provide this quantity of energy in a year. Because this energy would be radiation we would need to convert it to electrical energy to make a fair comparison. Using solar cells at, say, 5 per cent efficiency which is easily attainable with current technology, an area of desert of about 20 000 km² would be needed (a square with sides 140 km). This is not a feasible arrangement, though America has much more hot desert area than this, but is put forward as an indication of the plentiful nature of solar energy. As a practical energy source it has the disadvantages of being spread thinly over the Earth's surface (least intensely where space heating is most needed) and it is variable. Nevertheless public buildings and houses even in northern England during winter have been very successfully heated by solar radiation alone. The feasibility of canal barges and farm tractors being powered by electricity from panels of solar cells has been demonstrated.

Hydroelectric power is already well established in the world though it only supplies less than 10 per cent of the current energy production overall. In regions with appropriate conditions (for example, Norway) it produces almost all the electric power generated. The world's total generating capacity from hydroelectric sources even if fully used would not meet more than a quarter of the present energy demand, and the regions which have the greatest untapped potential have the greatest economic problems as far as large-scale high capital cost schemes are concerned.

Wind power potential is similar in many respects to solar energy. At Oklahoma State University the average wind energy was found to be 200 Wm^2 perpendicular to the wind direction which is of the same order as the value for solar radiation. However, because the supply is so variable we need to be able to store energy or in some way smooth out the fluctuations in power. Storing energy is technically extremely difficult though there are proposals for using the electricity generated to decompose water into hydrogen and oxygen and use the hydrogen as a fuel.

The two remaining energy sources are tidal power and geothermal power. Tides, caused basically by gravitational attractions between Earth, Moon and Sun, represent the movement of an enormous mass of water and hence an enormous amount of kinetic energy and are both predictable and non-polluting. At present the only working scheme is near St Malo in France where the average tidal height range is about 9 m in the estuary of the River Rance. The generating station now operating there has a working capacity of 240 MW comparable in output with a small fossil-fuel generating station. The total world capacity for tidal power has been estimated at about 9 million MW h per year.

Geothermal energy flows out as heat from the Earth's interior at a constant low rate but at some points on the Earth's crust the flow is well above average and it is at these points that this source may be tapped. The general average rate is only 0.06 W/m^2, but where there is a large circulation of deep underground water near a geological fault it is possible for the conditions to be just right for the production of superheated steam which appears as a geyser but could be used to drive turbines and generate electricity. At present, power is generated in plants in Italy, Iceland, Mexico, Japan, New Zealand, U.S.S.R., and the United States but the total energy produced is less than 1 per cent of the world production. The long-term use of the plants is governed by the amount of water that can be withdrawn and by the rate at which the systems are clogged by the salts and other materials forced up with the steam.

The future for energy supplies for the world must clearly involve a decreasing dependence upon fossil fuels but whether nuclear power will become a safe replacement on a world scale is open to question. Solar energy has obvious attractions but its utilisation seems to depend on a decentralisation of power generation (in contrast to the trends with nuclear power) to match the diffuse nature of the supply. There are other energy conversion techniques at early stages of development which might be relevant; these include nuclear fusion reactors, magnetohydrodynamic generators and fuel cells. From a technological aspect proposals for future sources can be divided into two groups—those which require high capital input, large-scale and complex technology, and on the other hand, those which can be made simply and on a small scale, and which need comparatively little capital. The very difficult decisions about future power supplies relate closely to the social and political issues concerning the degree of centralisation and the size of social units.

3.2 Material Resources

Throughout the greater part of man's history the demand for materials for shelter, clothing and tools was relatively slight because the technology and the energy sources (particularly the high-energy fuels) were not available to allow complex and large-scale undertakings. His shopping list of materials was modest; it could be conveniently divided into three groups; the organics, the inorganics, and water. The organic materials are the carbon-based compounds which were parts of living plants or animals and they are exemplified by wood, animal skins and cotton. The foods and the fuels came from the class of organic compounds too but they have already been dealt with. These materials follow the more rapid natural cycles; while demand rates were modest they were, both in principle and in practice, renewable, even if locally sources might at times be brought near to exhaustion. The inorganic materials follow natural cycles which are slow and dominated by geological processes; while in principle they are renewable, in practice the replacement rate is so slow that effectively they are non-renewable. Examples are stone, glass and ceramics and metals.

With the new fossil-fuel based technology of the industrial revolution the shopping list became more ambitious both in terms of quantity and range of materials sought. Two new groups of materials began to become important in the later stages; the synthetics and the composites. Both are more correctly products of manufacture which depend on more basic natural materials but both have great importance in the modern world and both pose problems of disposal and are difficult to return to their natural constituent cycles. The synthetics mainly come from fossil fuels and the obvious examples are the plastics while the composites are a much newer group which includes metals and non-metals reinforced by fibres of such materials as glass, carbon and boron.

If we try to estimate the extent and lifetime of the material resources of the earth we find that this is a very difficult and complex process which can yield a range of different answers to match a range of different sets of assumptions about future scenarios. Even if we did know the exact location and pattern of concentration of all the materials in the Earth's crust we would still not be able to say with certainty what the lifetime of a particular material will be. To attempt this we would need to make assumptions about the pattern of future demand, the development of new techniques, the availability of energy for extraction, transportation and processing and (because geological concentrations do not distribute themselves favourably across the national boundaries) the trading and power patterns of international politics. All these and many other factors, too, are interrelated in a complex manner, and are the subject of further discussion in chapters 5 and 6. Our treatment in this chapter merely investigates the consequences of simple patterns of resulting resource use. Yet it is important to make and specify assumptions and produce forecasts of lifetimes because decisions are being taken about long-term planning on materials supplies. Whenever such planning is undertaken we find too that the estimates do have a similarity despite quite different sets of assumptions. One author may produce an estimate for the lifetime for lead reserves as 15 years; another produces an estimate of 30 years; it may be mathematically sound to comment that one is twice the other but in terms of planning decisions it is clear that the lifetime is dangerously short. Rising population and *per capita* demand characterise the situation and with doubling times for both much shorter than a human lifespan a large percentage uncertainty in a short reserve life may be of little practical importance.

The most important geological aspects arise from the fact that, although most of the minerals we use can be found in very small quantities in most of the rocks of the Earth's crust, geological accidents in the remote past have resulted in concentrated deposits of chemical compounds which contain the materials we are interested in. These ore deposits are generally so much richer than ordinary crustal rock that the possibility of extracting the mineral that exists in trace quantities in ordinary granite is taken as negligible. The cost in energy terms of separating and replacing the unwanted crushed granite which would be present in the ratio of 2000:1 at least, and which would occupy a much larger volume than the original rock, is generally considered to rule this out as a possibility. The dominant source of metals will continue to be the ore deposits with metal concentrations several orders of magnitude above that of common rock. Even with rich deposits of ore, as mining continues, the depths and distances involved increase and, in many cases, the grade of the ore decreases; both factors increase the difficulty and the energy cost involved in winning the metal.

Using the best geological data available to them in 1970 the U.S. Bureau of Mines offered estimates of the major factors needed to make simple forecasts of lifetimes for some of the more important raw materials of the world.[1] Box 3–1 shows how for aluminium these lead to estimates which range from 31 years to 100 years depending on the assumptions made. Similar calculations for a further eighteen vital raw materials show that only four are likely to have a lifetime of more than 100 years. These calculations do not specify the changes in international trading patterns, price fluctuations, technological advances and other factors which are important. They merely specify consumption, discovery and extraction patterns and show what kind of timespan we are dealing with, and, what is not immmediately obvious until these preliminary studies are made, that optimistic assumptions about new discoveries and increased recycling make little difference to the life of the resource if demand continues to grow exponentially. If we refine the mathematical model further, to take into account other factors, and use a computer to produce a graphical representation of the changing situation we find that, in response to rising costs and technical problems, production does not continue to grow exponentially but falls off and after about 90 per cent of the reserves are extracted falls effectively to zero. Again the timespan involved is changed only slightly by this procedure.

Geographically the Earth's resources are not distributed evenly and the more developed industrialised countries depend heavily on supplies from less developed countries. The relationship between producer and consumer nations changes as reserves get smaller and particular countries find themselves in a position of power as they control the supplies needed by others to maintain their industrial pattern of society. Wars have been fought which have resulted in changes in this pattern of control; even though these changes may never have been publicly declared intentions the correlation between the involvement of the developed nations and the resulting changes in control of mineral resources is probably not accidental. An outcome of such a change which is disadvantageous to a particular nation leads to renewed searches for alternatives. Rarely does this lead to a long-term solution because the substitute material is also finite in quantity and often the energy cost is greater. Plastics may substitute for metals but they are themselves products of dwindling oil supplies. Recycling of materials has been practised by the poor through the ages, but now merits consideration by the rich. If the design of industrial products is based on the assumption of

Box 3–1   *The Future of Aluminium Supplies*

In 1970 it was estimated that the total world reserves of aluminium were 1.2×10^9 tonnes and that the 1970 rate of usage was 1.2×10^7 tonnes per year. If this rate of usage was continued until stocks were completely exhausted the reserves could be expected to last for:

$$\frac{1.2 \times 10^9}{1.2 \times 10^7} = 100 \text{ years.}$$

Over the past 10–20 years the rate of usage of aluminium has been growing and the growth rate in 1970 was estimated at 6.5 per cent per year. At this rate of exponential growth the rate of usage would double every 11 years. We can estimate a new lifetime for the aluminium reserves by a rough calculation along the following lines. Let the growth of usage continue exponentially without regard to the availability of the metal; in (a) the rate of use at each doubling interval is shown, and in (b) the average rate of usage for the doubling interval. In (c) the average rate has been multiplied by 11 to give the total used in that interval, and in (d) these totals are added cumulatively to give the total amount of the reserves used up to the year shown. (All figures are $\times 10^7$ tonnes)

	1970	1981	1992	2003	2014	2025	2036	2047
(a)	1.2	2.4	4.8	9.6	19.2	38.4	76.8	153.6
(b)	1.8	3.6	7.2	14.4	28.8	57.6	115.2	
(c)	19.8	39.6	79.2	158	317	634	1265	
(d)	0	19.8	59.4	138.6	297	604	1234	2500

Now the estimated reserves were only 120×10^7 tonnes so somewhere about the year 2000 they would run out completely. Suppose now that, with this prospect in view, and the price of aluminium soaring, geological surveys showed new reserves which meant that the total was in fact five times as large, or 600×10^7 tonnes. This remarkable increase would only delay the exhaustion of the new reserves until about the year 2025, a gain of 25 years.

Suppose the metal was from 1970 all completely recycled to meet the exponentially growing rate of usage, we can see from (a) that somewhere around the year 2042 the rate of usage is equal to the total reserves (120×10^7 tonnes) each year.

These estimates of lifetimes are based on mathematically simple patterns of resource usage. They give some idea of the periods of time involved and show very clearly that exponential growth cannot be sustained.

abundant supplies of the materials which seem most suitable on technological grounds alone substitution and recycling possibilities are generally diminished. Future design criteria could well include durability and ease of reclamation of raw materials.

Water has been discussed as a material resource in connection with food production but it has a vital role to play in industrial processes. It is a material which follows a comparatively rapid cycle and it has important properties such as its high capacity for storing heat, its good flow characteristics and its capacity for recycling which have led to its place as an essential resource for industrial production. Most developed nations are at the point where water supply engineers cannot see how continued growth of demand can continue to be met. To produce 1 tonne of steel we need to have available 100 tonnes of water, 1 tonne of paper requires 200 tonnes while to form 1 tonne of synthetic rubber (used as a substitute for a scarce material) we need 2400 tonnes of water. These quantities of water are of course used only briefly and flow back into the river or lake, rather than becoming part of the product itself. In this respect we cannot add demands together to obtain totals for a particular area because water is used and reused. In some of the major rivers of the industrialised nations the water which reaches the sea has had as many as 40 or 50 different users *en route*. When water is used in almost all cases some is lost by evaporation or transpiration, returning to the atmospheric part of the water cycle, or its quality is reduced by the addition of pollutants. There is of course a limit to the amount of water which can be drawn off by any particular user, which is less than the natural flow rate for the river (because other uses still require a continuous river flow) and must take into account seasonal variations and the minimum dependable flow. When the detailed analysis is completed we find that regions of the world just do not have a large enough dependable runoff flow to meet the projected increased demand for the year 2000.

3.3 Disposal and Pollution

The question of the repeated reuse of river water leads to the consideration of the third stage of the process with which this chapter is concerned and to the subject of pollution. When we have finished with the materials we have extracted and processed they are disposed of as waste and sooner or later find their way back into their natural cycles. Modern industrial waste is, however, concentrated as a result of the manufacturing process so that it is important to achieve dispersal and dilution. Samples of air and water have always, even before man appeared on the Earth, carried small amounts of other materials; air and water are the transportation agents for the faster stages of all natural cycles. What distinguishes the consequences of man's activities is either the quantity of the impurity present or the nature of the impurity. Pollution, which has the first characteristic—pollution of quantity—may arise from three different situations. In the first the natural level of a dangerous or even toxic impurity may be just below the danger level but the additional introduction of a further small amount may raise the concentration above this level (mercury and lead are examples of this situation). The second is a situation where mixing and hence dilution does not occur and a local hazard is created (oil spills and smog are examples). The third is where the impurity is harmless in itself and normally present but is introduced in quite large quantities and has the effect of disturbing the natural processes (nitrogen in fertilisers leading to eutrophication as discussed in 2.4, and waste heat). Pollution which is distinctive because of the fact that the pollutant is man-made and the natural systems are unable to accept the pollutant, forms the other distinctive group typified by the persistent synthetic pesticides. All these examples are the subject of case studies in section 3.3.1 where they receive more detailed attention.

As population increases and, at the same time, industrial activity is expanded, pollution must inevitably increase too. It may be possible to counter or minimise its effects but the likelihood is of an exponential increase. The evidence we have from pollution monitoring is that the pollution records show exponential growth at a rate slightly above that for population growth. It is important too to recognise the extent of our ignorance and uncertainty about its effects. We do not know what tolerance natural systems have to disturbance by pollutants nor are we able to say with certainty what are appropriate maximum permissible levels in the human body. We do not have sufficient knowledge of the patterns of distribution of pollutants across the globe but we do know that many are now globally distributed. We have not accumulated sufficient monitoring data to be able to make projections for many pollutants and natural systems have built-in delays so that we receive indications of serious levels of pollution too late to allow effective controls to be applied. These features will be evident in the studies of particular pollutants and pollution problems which follow.

3.3.1 *Pollution Case Studies*

(1) *Mercury*, though fundamentally toxic, has always been present in man's environment in such small concentrations that it has not presented a global hazard to health. Where it has been deliberately concentrated for some specific process it has presented a serious risk particularly to those who work

with it. Lewis Carroll's Mad Hatter in *Alice in Wonderland* (1865) was not mad but suffering from the poisoning observed among hatters (who, in the process of curing the felt, used mercury) which led to loss of control of the vision, movements and speech, interpreted then as signs of madness but now known to be symptomatic of mercury poisoning. This was, however, a specific occupational risk; since Carroll's time organic mercury compounds have been more generally observed in the air and water and have found their way into food chains. Some micro-organisms have the capacity to convert the element mercury into alkyl mercury compounds of which ethylmercury is the most toxic. We know that at the turn of the century mercury levels in fish were being recorded and increases noted but it was not until 1953, when a serious outbreak of mercury poisoning occurred at Minimata Bay in Japan, that there was a general awakening to danger.[2] A factory using mercuric oxide in producing acetaldehyde increased its output and, as a consequence, its discharge of waste mercury into the bay. Some time later sickness among local people became evident; there were 111 cases of serious illness (with 41 deaths). This was called at the time Minimata sickness, as though it was a local oddity, rather than a classic example of mercury poisoning. Later it became apparent that instead of diluting the mercury waste (the original intention in discharging it into the vast ocean) the water movements in the bay had kept it in shallow waters where, in the process of transfer through the food chain, it had become concentrated in the tissue of fish by a factor of several thousand times the concentration in the water. The diet of the people living around the bay was heavily dependent on local sea food. The response to the danger was not sufficiently rapid to prevent a similar incident in 1967 in Niigata City in which 26 serious cases, including 5 deaths, were recorded. The first recorded case attributed to seafoods bought in city shops appeared in 1971.

Mercury is introduced into the environment in a variety of ways, including the production of medical and dental preparations (broken thermometers released 7000 kg each year in Canada alone), the preparation of agricultural fungicidal treatments and in paints. Plants producing chlorine and caustic soda, and plants for pulp and paper normally use mercury, although since 1971 some countries have abandoned mercury in the pulp and paper industry where it is not essential. Mercury does find its way into the finished paper, however, and much paper still in existence was produced using mercury. The mercury found in coal and oil (from natural sources) and in paper and much domestic waste is released when the materials are burnt. Recent airborne surveys have shown that mercury is being discharged into the atmosphere in amounts which despite the low concentration, could be large by comparison even with mercury from industrial waterborne discharges. N.E. Cooke has estimated the world-wide release of mercury into the environment to be in excess of 10 000 tonnes annually.[3] We know now that mercury can persist for at least 100 years in coastal waters and that the effect of passage up the food chain is to increase the concentrations. Most developed countries monitor and control the level in marketed fish and seafoods but there is not sufficient knowledge of the effect of long-term exposure to small doses to be confident about maximum permissible levels and there seems to be enormous variation in sensitivity between individuals. Symptoms have been observed in individuals whose blood concentration was as low as 0.2 parts per million (p.p.m.) while individuals with 0.6 p.p.m. were classified as clinically healthy.[4]

(2) *Lead*, like mercury, is a highly toxic heavy metal present in the natural environment even before its use by man: like mercury it does not appear to be an essential trace element in human metabolism. It is estimated that this natural distribution of lead means a concentration in the atmosphere of about 0.0006 $\mu g/m^3$ while in the human bloodstream the level was 0.03 p.p.m. The rate of lead concentration has increased, especially over the past 50 years and has led to values as high as 71 $\mu g/m^3$ (for a peak traffic measurement on a Los Angeles freeway in 1966) and bloodstream levels for city dwellers of about 0.3 p.p.m. The level of lead in the atmosphere has thus increased many thousand times while the level in the blood has seen a tenfold increase. Even the general atmospheric levels in cites run at about 3–5 $\mu g/m^3$. The effects of lead poisoning on man are well known and, like mercury, it can cause serious damage to the central nervous system. It has even been suggested by S.C. Gilfillan that the decline of the Roman Empire might in part have been due to lead poisoning: the Romans lined their bronze kitchenware with lead to avoid the more obvious risk of copper poisoning and the evidence shows high levels of lead in the bones of upper-class Roman citizens.[5]

Lead reaches man through the food he eats (through lead deposits in the soil), the air he breathes (by direct absorption in the lungs), and the water he drinks. The comparative importance of these pathways is changing; 100 years ago the general population took in its lead through the digestive system and only workers in the lead industry and people living down wind from lead smelting plants received a major dose from airborne lead. In 1924 lead (tetraethyl lead) was introduced in petrol to improve car performance by making possible the use of higher compression ratios. Its use for this purpose in America alone rose from 400 tonnes in 1925 to 280 000 tonnes in 1969 although U.S. regulations have restricted its use since then and the quantity is no longer increasing. The U.S.S.R. never adopted leaded fuels and it is significant that measurements taken in Russia seem to be much lower than their equivalents in the United States or Britain. Children, because they are growing, handle heavy metal pollutants differently from adults and may show unremarkable blood levels while building up dangerous levels, over a period, in their bones. Mild lead poisoning can contribute to psychological and emotional disorders which are attributed to other causes particularly in deprived city centre children.

Mercury and lead are only two of the heavy metal pollutants that are extremely toxic and persistent and they are the two we know most about. Some, in contrast, are vital to living organisms in minute quantities, like zinc, copper, cobalt and selenium, but are highly toxic when this natural requirement is exceeded. All of them have been studied to some extent but there is a lot more that needs to be known before we can move from arbitrary to scientifically-based maximum levels and there is almost no information about the effect of two, or more, on the human body simultaneously.

(3) *Oil* has become the most important fuel for modern industrially-based society and the raw material for a wide range of industries. With world crude oil production running at about 2.5×10^9 tonnes each year, much of this originating thousands of miles from the point where it will be refined and used, it is not surprising that some will find its way into the natural environment. Oil production, transportation and use are heavily concentrated on the coastal regions and there shallower waters take the brunt of the pollution that results from oil losses. The high seas, however, are not immune; it is interesting that Thor Heyerdahl on his transatlantic journey in 1970 commented that he was rarely out of sight of lumps of tarry residue several centimetres in diameter. However, the coastal waters not only receive the bulk of the oil but are more seriously affected by pollution. We depend on these shallower waters for almost all our food from the sea; we also depend on these waters to absorb a large part of our industrial domestic and agricultural waste. Optimism about increasing food production by greater use of coastal waters seems to neglect the other demands made on them. It is estimated that at least 1 million tonnes of oil find their way into these waters each year while some observers put the figure as high as 3 million tonnes.[6] The spectacular oil spills from accidents like the *Torrey Canyon* shipwreck (1967, off the Cornish coast) and the major leakage from the Santa Barbara oil rig in 1969 amount to only about 10 per cent of the total; the remaining 90 per cent comes from normal activities involved in the transfer of fuel between ships, or ship and harbour, and the approved disposal of oil wastes from shore-based industries. Even more difficult to estimate is the amount that comes from fall-out of air-

borne hydrocarbons resulting from incomplete combustion but most observers estimate that this contributes a larger amount than the sources so far mentioned. Thus to comment that 1 part in 1000 of all the oil produced finds its way into the sea in the form of large hydrocarbon molecules is a safe but probably conservative estimate. Although spills from accidents involving large tankers are rare and contribute only a small part of the total oil load they do cause major local problems. The scale of the accident danger increases with the size of tanker. It is estimated that a quarter of all the oil transported in tankers passes through the English Channel. Supertankers of 400 000 tonnes deadweight are in one sense simple vessels in that their design is dominated by the need to carry as much oil as possible. The bridge is at the stern and they have only one propulsion screw. When a decision is made to stop the tanker the engines are shut down and it glides to a halt, but only at a point which is at least 15 km from the point where the decision was made. A collision which resulted in a loss of oil could easily produce a 50 000 tonne oil slick: in 1970 the direct cost of measures to deal with a slick of this size was between £250 000 and £350 000 and the fine imposed on the owners would have been unlikely to exeed £5000. Measures to deal with oil slicks are expensive and not particularly effective. The detergents used on the beaches after *Torrey Canyon* totalled about the same mass as the oil involved and did more damage to marine life, killing off some of the organisms that would have helped clear the oil.

Oil is an extremely complex mixture and the oil spills and leaks can involve very different proportions of the major components but almost all of them are toxic to marine life. The consequences can include direct killing, by asphyxiation or coating of filter-feeding animals (clams and oysters) and fish and seabirds. Subsequently the spread of dissolved or suspended toxic components, and the pesticides which are highly soluble in oil films, may lead to long-term disruption of food chains over a much wider area.

(4) *Smog*. This is a term originally used to describe a combination of smoke and fog but now more widely used to describe many serious conditions of visible air pollution. It occurs when the atmospheric conditions prevent dispersion and in fact lead to a concentration of pollutants over a period of time. Generally the atmosphere seems so large and turbulent a reservoir that a tall chimney and a slight wind will be enough to distribute smoke and the other products of burning so widely that they will not present a nuisance or a health hazard. This assumption has been proved to be wrong in a variety of particular situations but it is generally true that air pollutants are now distributed widely through the atmosphere; smog is observed at the North Pole

and the overall reflection–absorption–scattering characteristics of the Earth's atmosphere are measurably different now from 50 years ago. The pollutants are still with us even if we live in a clean air zone and our window-sills are cleaner than they were. Larger particles in the atmosphere will tend to fall out under gravity, unless the upcurrents are strong, and the general tendency of particles to grow by accretion (especially where they act as nuclei for the growth of water droplets and hail), does increase the likelihood of at least the larger particles reaching the Earth again. The wind strength governs how far they travel as well but a major factor is the vertical temperature gradient. Generally air temperature decreases with increase in height in the lower atmosphere, but not always. A pocket of air which starts to rise will expand and cool as it rises; if its fall in temperature still leaves it warmer than the surrounding air it will continue to rise and pollutants will be carried higher. If it becomes cooler than its surroundings it will sink and a stable condition will result with little or no vertical movement. It is possible in particular situations to have an atmospheric condition known as temperature inversion, where temperature increases with height. If this exists above a large industrial city with a high level of heat and smoke output when there is little surface wind a particularly unpleasant situation can result in which pollutants are trapped at ground level and accumulate. Any sunlight reaches only the top layer of smog and so makes the temperature inversion worse.

Serious incidents have arisen from such stable atmospheric conditions in many places; those in Donora, Pennsylvania, and in London are among the best known. In Donora, a small industrial town in a steep-sided valley, autumn conditions with temperature inversion in 1948 led to smoke and pollutants from various industrial plants (particularly zinc, sulphuric acid and steel plants) being trapped and building up in intensity over five days causing illness for half the inhabitants and death for twenty. In London in the winter of 1952 coal fires were the normal source of domestic space heating. In December the sulphur dioxide level in the atmosphere was double its normal value and thick yellow smoke combined with water droplets in fog to produce a lethal smog. After three days, visibility at ground level had fallen to one metre in the worst affected places and hospitals were choked with patients suffering from respiratory illnesses. About 4000 people died as a direct result of this incident (they were mainly very young or elderly) but large numbers of people suffered permanent damage to their lungs in addition. Los Angeles, Mexico City and Tokyo are regular sufferers from this kind of incident, though the pollutants trapped near ground level may differ from case to case.

There are several common pollutants which cause great damage to man. Carbon monoxide displaces oxygen in the haemoglobin in the blood preventing the supply of oxygen normally transported in the bloodstream reaching the cells which need it, so causing suffocation. Sulphur compounds (particularly sulphur dioxide) are contributors to respiratory diseases; they cause irritation in the lungs and, combining with water, create sulphuric acid and bronchitis and asthma sufferers are seriously affected. Nitrogen oxides operate in a similar way to carbon monoxide. The hydrocarbons and the pollutants which arise as particles, often from incomplete burning of fuel (or from processing materials like asbestos) are believed to be causes of cancer. All of these pollutants except the last arise specifically from exhaust emissions from cars and this forms a major urban source of air pollution. The evidence indicates that where they exist in combination their danger is greater than the sum of their individual dangers. It also seems to be possible for the action of solar radiation to lead to chemical reactions which produce from these pollutants new compounds which are even more damaging (photochemical smog).

(5) *Nitrogen* applied to farmland as a fertilising agent and subsequently washed out of the soil and into rivers and lakes was discussed in an earlier section on the techniques of agriculture (section 2.4) The promotion of growth in the water rather than on the land leads to undesirable effects and justifies the term pollution—but the nitrogen itself can be a health hazard in drinking water.

(6) *Waste heat* is a further example of this type of pollution situation which leads to a disturbance of a natural cycle. In chapter 1 the notions of energy conservation and the relatively low efficiencies achieved in most energy transformations were discussed. Above large cities and industrial complexes we already have convection currents due to local heating which result in local climatic anomalies; the annual mean temperature in such places is usually 0.6 to 1.2 °C above the comparable rural site. This effect is difficult to isolate because together with the waste heat we also have a higher concentration of pollutants and possibly of water vapour too. We know that over the 10 000 km basin of the Los Angeles district the rate at which heat is dissipated as a consequence of man's activities is currently about 5–6 per cent of the solar energy received on the same area and this fraction is increasing rapidly. In other more specific situations we have a more easily identified problem. Electricity generating stations draw water from rivers, lakes or estuaries for cooling purposes and then, after some considerable heat exchange with the surrounding air, discharge the warmer water back into its source usually

between 5 and 8 °C above its original temperature. As temperature rises it is generally true that activity and growth rates for water plants and animals increases. However, it is also true that warmer water cannot hold so much dissolved oxygen so that we have an increased oxygen need matched by a diminished supply. The water is less able to cope with organic waste and, in consequence of the deterioration in their environment, particular species of plant and animal life may be eliminated from that stretch of water. A small temperature rise and specially designed ponds can mean a more favourable situation for some species and lead to better prospects for fish farming. A large temperature rise can kill fish outright. In Britain cooling towers which take away a lot of the waste heat by evaporation are commonly used to lessen the thermal load on rivers. Nuclear generating plants are less efficient than fossil-fuel plants and will increase this thermal load considerably. Currently the production of electric power is growing exponentially at about 5 per cent, doubling every 14 years.

The earth maintains an overall thermal balance and reradiates energy at the same rate as it receives it from the sun. As it is neither gaining nor losing energy in this condition its temperature remains constant. During Earth's history there have been major changes in the surface, both in the atmosphere and in the land coverage, which will have changed the equilibrium conditions slightly, leading to a balanced condition but at a slightly different temperature. Over very recent history man's activity, especially the burning of fossil fuels, has led to an unbalanced situation; the Earth would as a consequence tend to get warmer. At present the rate of non-equilibrium heat production by man is only about 1 in 21 000 of the solar radiation input and the temperature rise that would result from this small inbalance is too small to detect. If we consider how this might grow, and we assume that the growth rate for dissipation of energy is about 5 per cent per year, then, using the 5 per cent figure (and a doubling period of 14 years) the ratio would become 1 to 100 in about 100 years time. A further 28 years would reduce this ratio to 1 to 20, which corresponds to a 3 °C rise in temperature. Such a rise would mean that the ice caps would melt and much of the present dry land would be flooded. These calculations are most unlikely to be a prediction of the actual situation in, say, A.D. 2120, firstly because this is a situation which emerges gradually and policies could change, but secondly because it means an increase of about 1000 to 1 in the production of energy over the present rate which presents major difficulties of achievement.

(7) *The synthetic organic pesticides* are typical of the man-made pollutants which have been fed into the natural cycles over the last forty years and an

account of their harmful effects has been given in section 2.4 when discussing pesticides. Many cases of specific use of insecticides and herbicides have now been carefully researched and detailed accounts are available. One which may prove significant in world history is the use of herbicides as weapons of war, as exemplified in the defoliation of Vietnam which was intended to reduce the natural cover and make control of the area easier. Herbicides were applied at an average of thirteen times the strength normally recommended for agricultural use and one of the herbicides was Pecloram, which is prohibited as far as use on any crop-bearing land in the United States is concerned. The complete or partial destruction of some 20 000 km^2 of forest and farmland with poisons whose effects will still be evident for several years may not have been widely publicised, but it does represent deliberate destruction of a living environment on a scale and at a rate never before known to man.

Only a very few examples of pollution can be studied in a book of this size, but it would be wrong to exclude two further case studies which do not quite fit into the classifications given in the introductory section 3.3; both refer to future hazards and both are the subject of controversy.

(8) *Disturbance of the Stratosphere.* There are situations where the extent of our uncertainty is great because the problem is complex, several pollution mechanisms are operating simultaneously and possibly interacting, and monitoring in the early stages is difficult. When we consider the risk of damage to the stratosphere by the release of gases directly from the jetpipes of supersonic aircraft (S.S.T.) and indirectly from our discharge of persistent chemical pollutants which find their way up through the atmosphere, we are in such a situation. The stratosphere is the layer of the atmosphere that lies between 12 and 50 km above the Earth and that does the bulk of the work of shielding the Earth from the harmful radiations that come from the Sun. It differs from the lower atmosphere in that its temperature pattern is reversed—temperature increases with altitude—causing a temperature inversion situation which, as we noted when discussing smog, is a stable situation. Any pollutant reaching the stratosphere stays in this zone and will not rise above it but it will be distributed globally because wind speeds are generally high. This stability, as far as the gases upon which we depend for this shielding action are concerned, is an essential characteristic. The higher temperature at the top of the stratosphere is a consequence of the shielding action and in particular the production of ozone. Ultraviolet radiation splits oxygen molecules (O_2) into the two oxygen atoms, which then combine into

ozone (O_3) with an accompanying absorption of heat.

The advent of high-speed, high-flying aircraft, like Concorde and its successors in the projected S.S.T. range, led to the suggestion that nitrogenous oxides, sulphur dioxide and water vapours which would certainly be introduced into the stratosphere could reduce the formation of ozone, leading to an increase in ultraviolet radiation at ground level, which could be extremely harmful, and to climatic changes. There is still insufficient agreement amongst scientists about the extent of this danger although it is agreed that a risk could exist when fleets of S.S.T. are in service. The report of the Federal Aviation Administration in March 1975 recommended allowing Concorde to operate over America, but included serious warnings about future expansion of stratosphere contamination. More recently the gases which act as propellants in aerosol sprays have been found to diffuse into the stratosphere and have a similar effect on the ozone layer; in fact the only way these compounds can be broken down is under conditions of high energy ultraviolet radiation and it seems possible that after a considerable period of diffusing through the lower levels of the atmosphere they reach a level where they also have the effect of depleting the ozone.

(9) *Nuclear Radiation.* The natural environment of man has always included some ionising radiation (referred to as natural background) and this, like any other radioactive effect of its kind, can do damage to human tissue. We have grown accustomed to living with this as a fact of life although its explanation is only very recent. What is new in our situation is that over the past 75 years man has introduced a large number of new radioactive sources. The effect of exposure to radiation includes damage to the person exposed, which for a larger dose could mean death within hours or days, and damage to future generations because the reproductive cells are particularly vulnerable and defects are transmitted genetically. The major additions to the radiation load have come from two sources, the fall-out of radioactive materials from the testing of nuclear weapons and the use of radiation, particularly X rays, in medicine. The use of radioactive materials has extended into many fields including agriculture and engineering, as well as medicine, but controls and safeguards have been maintained and accidents are rare and well contained (at least as far as the developing countries are concerned). The problem with radioactive materials is that they go on emitting radiation and they cannot be switched off. For some the process of emission goes on at dangerous levels for very long periods. The expression 'half-life' is used to indicate the duration but this needs some explanation. Radioactive materials decrease in activity in what could be described as the reverse of exponential growth—that is,

exponential decay. Just as we were able to talk of a 5 per cent per year growth rate leading to a doubling period of 14 years so with radioactive decay we speak of the period in which the level of radioactivity drops to half its initial value as the half-life. Many half-lives are very long indeed, and some of the materials at present used in hospitals and subsequently disposed of as waste still have a long period during which they can only be handled under special provision by skilled personnel. The half-life is merely the time taken for the radioactivity to decay to half its original level—this level could still be highly dangerous to man. The sheer bulk of low-activity waste from these users is now becoming a problem. At present the European nations co-operate in a scheme in which the materials are dumped, in drums, into a deep part of the North-east Atlantic.

The use of nuclear power in generating electricity has meant that more highly radioactive materials are being processed, transported, burnt in reactors, and finally disposed of as waste, than ever before. Accidents at nuclear reactors have already occurred but have been very slight and major accidents are extremely unlikely but not impossible. The probability becomes greater as the number increases and as reactors are built in countries where the government has only a precarious hold on affairs and where sabotage or civil war could occur. Situations like this increase the risk associated with the transporting of nuclear fuels and materials. The routine disposal of waste from reactors involves less bulk than the materials used in agriculture and medicine but the level of radioactivity is extremely high and the half-lives are longer. In Britain and the United States spent fuel rods are dissolved in acid and after further treatment to reduce the bulk the solution is stored in large tanks. The material is still extremely radioactive and generates heat in large amounts which have to be absorbed by heat exchangers to stop the liquids boiling. At Windscale in Cumbria the tanks have double skins of stainless steel over a centimetre thick and an outer cladding of concrete of 1.5 m. They cost over £2 million each to build and will need to be tended with great care for 700 years because until then the material will be too radioactive to handle. After this a new simpler storage tank will probably suffice for the remaining 10 000 years since for most of this time the level of activity will be much lower, but the materials themselves will remain highly poisonous. The materials will remain dangerous for a period which must be reckoned in millions of years since the half-life of neptunium, one of the normal components of the waste, is 2 million years. New, better storage systems are being sought. We can be thankful that these most dangerous of all pollutants are at present 'pollutants contained' unlike the others already studied. The stored material is continuously growing in quantity and will need the

vigilance, skill and integrity of 2800 generations of specialist workers to prevent it becoming a pollutant 'at large'.

Summary

We have examined the diversion of materials from their normal pathways in
the natural cycles into the extraction–manufacture and use–disposal loop
which characterises man's activities and have seen the key role which energy
has in this process. The growth of demands and the limitations of resources
available to meet them have been discussed in the light of the relatively short
lifetimes of many of the materials we currently use, and we have seen the indications that the age of industrial technology could well be a short episode in
the whole span of human history. We have also seen how difficult it can be to
identify a particular resource or environmental problem in isolation; the inter-
connections are usually most obvious and important. Many of the solutions
offered for the problems of shortages of materials presuppose abundant
supplies of cheap energy in the form of simple fuels, and many substitute
materials are themselves chemical by-products of these same fuels; the
proposed solution to almost every problem of this kind assumes the
availability of fuels which are demonstrably shortage materials themselves.
To propose increasing the harvest of food resources by greater yields from
the shallow coastal waters presupposes that we will stop, or minimise, over-
use of these waters as a convenient sink for the unpleasant wastes of our industrial society.

Pollution, viewed simply as a technical problem, does not pose great
difficulties; assuming no restrictions on cash flow and energy, and the easy
acceptance of restrictions on our personal freedom and our industrial activity, it could be reduced dramatically. We would still have the legacy of the
persistent pollutants, already well established in natural cycles, which could
only be eliminated gradually. The problem is not a simple technical problem
precisely because these assumptions are unlikely to be valid.

CHAPTER 4

The Environmental Debate

The first three chapters have discussed the ecological principles and the major factors involved in any study of the relationship between man and his environment. In this chapter we shall attempt to show the gradual development of what could prove to be the most crucial debate in the history of humanity and the present position and arguments of the leading figures who are contributing to it. The debate centres around the environmental crisis and the extreme positions are easily summarised. On the one hand there are those who contend that there is a crisis, now; that the threat to man's survival is so severe that only dramatic, urgent and drastic remedies can give any hope, and that damage to the environment is already interfering with the life-support system on a global scale and will lead to a breakdown of society, and to famine and disease within the lifetime of children alive today. At the other extreme there are those who argue that this is a gross and pessimistic misreading of the signs; we have time to remedy the errors of our technological development (which are nowhere near as crucial as the environmentalists suggest) and there are reserves of materials and energy, and a resilience in our life-support system, which give cause for optimism about the future. The environment and the threat of an environmental crisis became matters of public debate during the mid-sixties; before this isolated prophetic voices gained some public attention but, because they were isolated and because they were giving warning of the dangers of the future at a time when the natural environment did not seem to be showing obvious signs of damage, their prophetic words had little lasting impact. It would be quite wrong to allow the impression that environmental damage is a unique feature of the last 200 years, or that primitive man was, in some subtle way, instinctively attuned to his natural environment so that he only damaged it when he made errors through lack of skill or knowledge. There must have been enormous damage to the fertility of the soil, destruction of forests and the Pleistocene era must have seen the extermination of complete species of animals over what are comparatively long periods of time. The rates of his operations were probably much slower than those we see today but we are left with the impression that the life style of early man contained the same tendencies to destruction but that he lacked the opportunity and the techniques to match our present achievements.

4.1 The Development of the Debate

Before the industrial revolution specific contributions to the debate in the terms described above are understandably rare, though there has never been a shortage of contributions to the more general debate about the nature and behaviour of man, and his relation to the natural world, and to other men, which are distinctly relevant to our subject. The global view is a modern privilege however; the scientific and technical developments which brought the ability to gather data, communicate and manipulate it with speed and facility has made possible studies which could not have been considered fifty years ago. One voice which does stand out in the early years of this period is that of Malthus.

In the final years of the eighteenth century, when the population of Western Europe was well into its greatest ever growth period, a young British parson and scholar, Thomas Robert Malthus, published the first of his two essays *On the Principle of Population* (1798). This essay, so often misquoted and misunderstood since, caused consternation amongst his contemporaries with its forecast of imminent catastrophe arising directly from the pressure of population. Malthus, a Cambridge mathematics graduate, started his argument from the dual premise that man needs food to survive, and that human sexual passion is likely to be a constant feature throughout man's history. He then reasoned that where food was available, populations would tend to grow (unless there were very effective checks) and that the power of population growth was inevitably greater than the capacity for food production. Malthus put it that while food production grows arithmetically, population grows geometrically. If then population rises it must inevitably meet some check and the check may be the result of deliberate choice (moral restraint in his terms) or 'misery and vice' by which he meant war, famine, disease, hardship and extra-marital intercourse and the use of contraceptives.

Malthus was wrong in this time scale: he could not have foreseen the developments in agricultural practice which have kept the *per capita* food production at a more satisfactory level than he would have thought possible. He could not have foreseen the great medical revolution which has brought death rates down so dramatically and as a consequence increased the power of population. Although these two factors are in opposition they have had the overall effect of delaying the catastrophe he visualised. We would not find many people today who would be happy about his elitist assumptions or his definition of vice and his total rejection of contraceptive devices, but his overall case seems to stand. We may have bought time with our ingenuity but the capacity of the Earth to produce subsistence still remains finite and is in fact threatened in the long term by the short-term techniques which disturb ecosystems. Malthus saw that self-restraint was quite possible and stressed the responsible role of governments in controlling populations; he was not however, able to find much ground for optimism and his essays are still frequently attacked as presenting a 'dismal theorem' while 'Malthusian' is seen as a term of abuse.

In the period between 1803, when Malthus's second essay on population was published, and the mid 1960s, when there was a flurry of contributions to the debate, relatively little was written which had any formative effect on the debate as we know it. During this period the pure scientists were working in increasing isolation, making enormous strides in their specialist areas but understandably sparing little effort for integrating their work. It was from the ranks of the geographers that some very valuable contributions were made. Outstanding in the published material is the work of George Perkins March, at one time United States ambassador to Italy, whose book *Man and Nature, Physical Geography as Modified by Man* (1864) was a pioneering work in this field. He revised his book several years later as *Earth Modified by Human Action* and his influence was powerful in helping the campaign in America to conserve natural life and beauty. The conservation movement, which was active and effective in America many years before it gained government and public support in Britain, was the major contributor to the debate during this period. Conservation is not the same as global ecology but is contributory to it: we may wish to fight against the destruction of a woodland because we see it as an amenity, as a thing of beauty, while the ecologist is able to go further and show that we are destroying something on which we depend for survival. It is a matter of opinion as to what constitutes scenic beauty and what changes made by man enhance a natural setting but there is little doubt that whilst the practical work of the conservationists may have to be seen as campaigning on a limited front for a particular place, building, plant or animal, the overall effect was to anticipate a broader campaign to protect the ecosystem as a whole.

In the years immediately following the Second World War it was natural to think about the future of man, and a spirit of optimism prevailed. Yet a warning voice in that setting came from several American writers who saw the dangers of a situation in which growth and development were not matched with understanding and care for the land. William Vogt's *The Road to Survival* (1948) and Fairfield Osborn's *Our Plundered Planet* (1948) were both received in America, where they were first published, as alarmist, emotional attacks on the agricultural practice of the time. We are now sufficiently informed to see that their early warnings were justified. Vogt wrote *People, Challenge to Survival* (1960) and Osborn *The Limits to the Earth* (1953) in the years that followed and these books have obviously had a very considerable influence on those who joined the debate when much more public attention and concern had been drawn to the environment. One further major source from the fifties was the collection of symposium papers edited by William J. Thomas, *Man's Role in Changing the Face of the Earth* (1956), which included contributions from Lewis Mumford, Teilhard de Chardin and Hermann von Wissman and reviews the history of man's concern about his impact on his natural surroundings.

4.1.1 *Commoner and Ehrlich*

The two giants of the American scene whose output has had a major effect on the ecological debate over the past 10–15 years are Barry Commoner and Paul Ehrlich. Both are leading University biologists and both entered the arena of debate with vigour: it is usually these two men who have drawn the criticism that they are too ready to cry doom. Both are prepared to campaign and to be outspoken because both see the crisis as upon us now, not in the future, and are aware of the traditional inertia of public opinion and government. It is an oversimplification to say that they differ in that Commoner sees technology as the root cause of the crisis while Ehrlich blames population pressure but their more public disagreements have given this impression.

Barry Commoner is Professor of Plant Physiology and Director of the Centre for the Biology of Natural Systems at Washington University, and a director of the American Association for the Advancement of Science. He has a tendency to look for ways of expressing quantitatively the ecological features he is concerned with and has related the deterioration of the environment to the rate of production of goods in a mathematical analysis which forms the foundation of his more recent published arguments. He wrote in a more descriptive manner in his most famous book, *The Closing Circle* (1963), but the theme is the same. The root cause of environmental degradation is that the technologies we have used over the past 75 years have been ecologically faulty and have been used long before we were aware of their ecological effects. The consequence of this degradation is seen in the stress on our life-support system and the destruction of our biological capital. The actual level of damage is now so high that it presents a threat to continued development and our priorities are to repair the damage already done and replace present technologies with ones that are ecologically wiser. (In box 4–1, a typical quantitative relationship between population, affluence and technology is shown.) He comments that the new technology is an economic success but only because it is an ecological failure and that

> We are concerned not with some fault in technology which is only coincident to its value, but with a failure that results from its basic *success* in industrial and agricultural production. If the ecological failure of modern technology is due to its success in accomplishing what it sets out to do, then the fault lies in its *aims*.[1]

Paul Ehrlich, Professsor of Biology at Stanford University, with a particular interest in entomology, population dynamics and evolution, has a fellowship at the Center for the Study of Democratic Institutions and, like Commoner, he is ready to join the debate at any point and is a prolific writer and speaker. He has been most unhappy about the way Commoner has been able to reduce population growth to a very small factor in the overall problem. He calls this one-dimensional ecology and argues that taking a base in 1900 and only treating the 75 years which follow may have made the data more accessible and reliable but it has also hidden the truth that serious ecological impact began 10 000 years before when man moved into settled agriculture. For Ehrlich the sheer pressure of numbers, the inertia of population growth, the social consequences of the threat of famine and disease are likely to be the significant aspects of our impending eco-catastrophe. The space which he and his wife devote to these aspects in their major work, *Population, Resources, Environment* (1972) amounts to half the contents of the book. He too indicts misguided technology and refers frequently to limits to other growths than population alone.

> Population control is absolutely essential if the problems now facing mankind are to be solved. *It is not, however, a panacea.* If population growth were halted immediately, virtually all other human problems—poverty, racial tension, urban blight, environmental decay, warfare—would remain. The situation is best summarised in the statement: whatever your cause, its a lost cause without population control.

Ehrlich goes further into discussion of the social and technical activity which he sees as vital for the survival of mankind than Commoner and devotes a

Box 4–1 Barry Commoner's Evaluation of the Environmental Costs of Economic Growth

Commoner sets up an index of environmental impact (I) which, typically is the amount of a pollutant which is introduced into the environment in one year. He then considers factors which relate to this impact, which he lists as affluence, technology and population. By *affluence* he means the *per capita* production of the goods which are associated with the pollutant studied. The *population* is the simple number of people involved and *technology* is the amount of pollutant per unit of production. He expresses the affluence and technology factors in more general terms as economic good divided by population (affluence) and environmental impact divided by economic good (technology). He then says that the environmental impact

$$I = \text{population} \times \text{affluence} \times \text{technology}$$

which follows from his definitions as

$$I = \text{population} \times \frac{\text{production}}{\text{population}} \times \frac{\text{pollutant}}{\text{production}}$$
$$= \text{pollutant}$$

Taking the example of fertiliser nitrogen (see sub-section 3.1.2) as a pollutant linked with food production (which brings both benefit and environmental damage) and using figures for the United States for 1949: crop production index 81 units, population 150×10^6 and nitrogen used in fertilisers 0.914×10^6 tonnes. These figures yield an affluence index of 0.54×10^{-6} and a technology index of 11×10^3. Carrying out the same calculations for 1968 we find that the affluence index is now 0.60×10^{-6} (an increase of 11 per cent) but the technology index is 57×10^3 (an increase of 405 per cent) and that during this period the population has increased by only 34 per cent. The environmental impact has leapt to 6.8×10^6 tonnes (an increase of 648 per cent).

From these calculations and a vast number of similar ones for other environmental impacts he concludes that the very large (405 per cent in the example given) increase in the technology factor is far more important than the small (34 per cent) rise in population as far as its contribution to the environmental damage is concerned. The environmental impact (648 per cent) is seen to be enormously larger than the benefit (11 per cent) in almost all the examples he studies.

(A much more complete development of this argument can be seen in Barry Commoner's 'The Environmental Cost of Economic Growth', in *Chemistry in Britain*, **8**, (2) (1972) pp. 52–65, from which the material of this box was taken.)

large part of his writing and speaking to proposals for change in the United States and for fighting famine and malnutrition in the Third World countries. The Ehrlichs comment rather sadly:

> It is unfortunate that at the time of the greatest crisis the United States and the world have ever faced, many Americans, especially the young, have given up hope that the government can be modernised and changed in direction through the functioning of the elective process.

Later in the same passage they speak of the need to support those politicians who do take some kind of stand and show some awareness and concern, and they conclude thus:

> The world cannot, in its present critical state, be saved by merely tearing down old institutions, even if rational plans existed for constructing better ones from the ruins. We simply do not have the time. Either we will succeed by bending old institutions or we will succumb to disaster. Considering the potential rewards and consequences we see no choice but to make an effort to modernise the system. It may be necessary to organise a new political party with an ecological outlook and national and international orientation to provide an alternative to the present parties with their local and parochial interests. The environmental issue may well provide the basis for this.[2]

The first edition of Barry Commoner's *The Closing Circle* appeared in 1963, and the Ehrlichs' *Population, Resources, Environment* in 1970. During these intervening years there had been growing attention given to the environmental crisis in the newspapers and on radio and television. The reaction against popular presentations was to label those who presented them as doommongers and to make comparison with hell-fire preachers who foretell the end of the world. The specific opposition to the case that there is an environmental crisis now was, and still is, largely based upon a blend of criticisms and articles of (optimistic) belief.

The criticisms of the case presented included doubts about the ability of any commentator (or team of commentators) to have data on, or knowledge of, the whole environment, in all its complexity, at a level of completeness and accuracy which justifies the radical conclusions reached. This criticism was often substantiated by particular examples of errors of data or relationships. Rachel Carson can be shown, now, to have included some erroneous data in *Silent Spring* (1963) and she certainly wrote outside her acknowledged specialist field. The book was attacked by specialist scientists and others but her overall conclusion—that D.D.T. and related pesticides not only kill insect pests but many other organisms as well, and endanger the life-support system as a whole—reached in 1962, has since been repeatedly confirmed by the research in this field despite the errors this research has shown. In many senses the criticism is acknowledged as valid by the environmentalists but the counter-comment is also made that their analysis may represent by far the best picture available, and that decisions do have to be made. The decision to continue with an existing policy is rarely backed by complete or accurate data and it is easy to sense an assumption that until a perfect case for change is presented, it is not necessary to make any change; proposals for reform are subject to greater scrutiny than reasons for accepting status quo.

The 'articles of faith' accusation is used by both sides in the environmental debate: Ehrlich only sees what Ehrlich wants to see, because he is already convinced of a radical environmentalist case though his data does not necessarily lead others to the same conclusion. Ehrlich's case has been put, earlier in this chapter, and it is worth considering the articles of faith underlying the opposition. One is that no problem has been revealed which cannot be dealt with by a more skilful use of science and technology; another is that economic growth is necessary to provide resources and goods in order that basic needs may be met and people can give attention to such matters as environmental degradation and the quality of life. A more subtle assumption seems to be that any vigorous campaigner for environmental issues has ulterior motives: he is suspected of having a desire to ensure that his class or nation suffers least in the future if things are going to get at all difficult. This suspicion is often found among Third World commentators and it comes as no surprise; after years of frustration caused by the growing gap between rich and poor, the environmentalists seem to be saying to them that, after all, there is not as much available as was at first thought, and so it would be unwise to go for industrial development and you, unlike us, had better get your growing population under control to reduce the pressure on what is available.

A misunderstanding also marred the quality of the debate which centred around statements made about the future. The words 'prediction', 'forecast', and 'projection' were used as though they were interchangeable. Prediction is the least scientific of the three words and literally means saying, before the event, what will happen. Forecast implies, when used with caution, that we have cast forward the trends and values which are evident at present, and as a result, we are led to the conclusion that a particular event is likely to occur. Projection implies a simpler and more scientifically precise statement—if this (stated) trend continues without change then this event will occur. The projection makes explicit the assumptions underlying the statement and these can be discussed and changed, if the evidence indicates this to be of value, and a further projection made. The last two are frequently interchanged; the first has little place in sound debate. Two major publications in 1972 must be seen in the light of this distinction.

4.1.2 *'The Limits to Growth' and 'A Blueprint for Survival'*

In February 1972 the report of the Club of Rome's project on the predicament of mankind appeared in the bookshops and 12 000 complimentary copies were distributed. Entitled *The Limits to Growth*, it was the product of co-operation between a group of international consultants, under the chairmanship of Aurelio Peccei, and an international team of scientists and computer experts at the Massachusetts Institute of Technology who had developed a new approach to the computer modelling of complex problems. The Club of Rome had begun as a meeting of about thirty individuals from about ten countries in Rome in 1968, and under Peccei's chairmanship grew to about seventy-five over the following two years; its members are all eminent consultants or educators and while none holds government office, all wield considerable influence. Aurelio Peccei himself is a major figure in the Fiat and Olivetti concerns, chairman of Italconsult and of the Committee for Atlantic Economic Cooperation. From these meetings had come an identification of the 'world problematique', the complex web of technical, economic, social and political problems which face all countries in the world and which interact with each other in such a way as to make isolated analyses of individual problems of only limited value.

In 1970 Professor Jay Forrester of M.I.T. presented a technique of mathematical modelling which would facilitate an exploration of these interactions using a computer. Forrester was already a leading expert on this 'systems dynamics' approach for analysing the behaviour of complex engineering and social systems, and he was joined by Professor Dennis Meadows and a small team initially supported by funds from the Ford, and later the Volkswagen Foundations. The team examined five chosen basic quantities whose levels indicate best the state of the world system—population, pollution, natural resources, agricultural capital (or output) and industrial capital (or output). They then set out to establish levels (often using 1970 values as a basis) and rates of flow together with the feedback mechanisms which seemed to describe their interrelations best. Feedback loops are more satisfactory devices for describing a responsive system than cause–effect links. If, for example, I am riding a bicycle, I can turn the handlebars and this has the effect of changing the direction of my path; however, I watch this new path and use this as a signal to govern the amount of turning I need, at any time, on the handlebars. This notion of a feedback loop was basic to the design of the world model. The team's practical approach was to determine by discussion with specialists the kind of behaviour the system had shown and was showing at the time, and then to model this mathematically until a fairly complete model was established. This was then run through the computer to establish a graphical presentation of the behaviour over the period 1900–2100. The implications and the merits of this presentation would be discussed and new models produced from this feedback. A typical computer printout is shown in figure 9 with explanatory com-

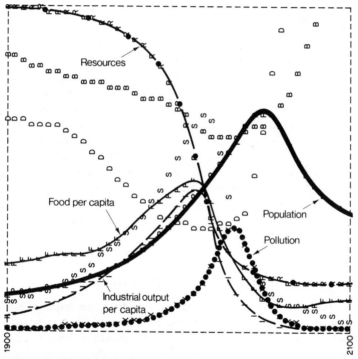

Figure 9 World Model Standard Run. The 'standard' world model run assumes no major change in the physical, economic or social relationships that have historically governed the development of the world system. All variables plotted here follow historical values from 1900 to 1970. Food, industrial output and population grow exponentially until the rapidly diminishing resource base forces a slowdown in industrial growth. Because of natural delays in the system, both population and pollution continue to increase for some time after the peak of industrialisation. Population growth is finally halted by a rise in the death rate owing to decreased food and medical services. (Computer printout and commentary reproduced from Dennis Meadows *et al.*, *The Limits to Growth*, 1972, by permission of Pan Books Ltd.)

ment. Some of the printouts which eventually emerged showed unexpected behaviour but the general conclusion was no surprise. The group shared an initial view that it was important to work on a global scale but, more important, that the world was finite and that growth of the five quantities chosen (for example) could not continue indefinitely. They were looking for the behaviour patterns that would result when these limits were approached and, in particular, seeking a smooth transition from a growth to an equilibrium condition in which population and the quality of the environment settled to stable and sustainable levels. The passage which follows shows this thesis well.

> There may be much disagreement with the statement that population and capital growth must stop *soon*. But virtually no one will argue that material growth on this planet can go on forever. At this point in man's history, the choice posed above is still available in almost every sphere of human activity. Man can still choose his limits and stop when he pleases by weakening some of the strong pressures that cause capital and population growth, or by instituting counterpressures, or both. Such counterpressures will probably not be entirely pleasant. They will certainly involve profound changes in the social and economic structures that have been deeply impressed into human culture by centuries of growth. The alternative is to wait until the price of technology becomes more than society can pay, or until the side-effects of technology suppress growth themselves, or until problems arise that have no technical solutions. At any of those points the choice of limits will be gone. Growth will be stopped by pressures that are not of human choosing, and that, as the world model suggests, may be very much worse than those which society might choose for itself.[3]

The report received both support and criticism. From the developed countries, economists and politicians criticised the economic thinking involved, especially the desirability of non-growth (or a steady-state economy), and the whole notion of limits to growth was under attack. Jeremy Bray, Fabian, former Labour M.P. and member of the Club of Rome, challenged the basic idea of the existence of physical limits, and spoke with confidence of the possibility for man to conduct 'unlimited technological migration', moving on from one technology to the next as the need arose. One year after the publication of *The Limits to Growth* the Science Policy Research Unit at Sussex University published a detailed critique entitled *Thinking about the Future*. Using its own computer facilities to substantiate objections to the model, the Sussex team showed that quite small changes in some critical relationships and rates of flow led to quite different outcomes. From the Third World came a different kind of criticism; this centred, not on the technicalities, but on the motives of the Club of Rome. In this view, the whole exercise was seen as designed to make easier the exploitation of resources by multinational companies and to ensure the perpetuation of an existing unjust distribution of wealth and power. The second round of this battle was begun in 1975 with the publication of the second report for the Club of Rome (entitled *Mankind at the Turning Point*), based on a computer world model and promoting the same basic propositions as *The Limits to Growth*, but carrying the analysis further and dealing with objections raised by critics of the 1972 report.

The January 1972 issue of the *Ecologist* magazine was entirely devoted to the presentation of *A Blueprint for Survival*, jointly written by Edward Goldsmith, Michael Allaby, Robert Allen, John Daroll and Sam Lawrence.

The five authors had seen, and were able to incorporate, the M.I.T. material which was published eight weeks later as *The Limits to Growth*. The *Ecologist* statement, supported by thirty-six leading British scholars, went beyond analysis of the physical and economic aspects; it presented a new philosophy about the kind of society that could be sustained for the indefinite future, and put forward a comprehensive programme for bringing about this social order. Its author saw *A Blueprint for Survival* (which was soon published as a book by Penguin) as heralding a new Movement for Survival, which never actually came to birth, though the organisations which were to form its main support are still extremely active in this field. It was envisaged that this movement, armed with *A Blueprint*, would influence governments—in particular that of Britain—to move towards a stable society. *Blueprint* took over where *Limits* stopped. Aurelio Peccei expressed support for the diagnosis in the M.I.T. report, but saw it as basically optimistic because it did not give sufficient attention to the social and political limits to growth, and because he foresaw the probability of a much earlier collapse or breakdown of social structures. *A Blueprint for Survival* makes specific diagnostic comments in this area, but also gives a strategy for change.

> ... The principal conditions of a stable society—one that to all intents and purposes can be sustained indefinitely while giving optimum satisfaction to its members—are (1) minimum disruption of ecological processes; (2) maximum conservation of materials and energy—or an economy of stock rather than flow; (3) a population in which recruitment equals loss; and (4) a social system in which the individual can enjoy, rather than feel restricted by, the first three conditions.

It then goes on to list seven operations which would be required to effect these changes and to propose a scheme for controlling and orchestrating them.

The achievement of these four conditions will require controlled and well-orchestrated change on numerous fronts and this change will probably occur through seven operations: (1) a control operation whereby environmental disruption is reduced as much as possible by technical means; (2) a freeze operation, in which present trends are halted; (3) a systemic substitution by which the most dangerous components of these trends are replaced by technological susbtitutes, whose effect is less deleterious in the short-term, but over the long-term will be increasingly ineffective; (4) systemic substitution, by which these technological substitutes are replaced by 'natural' or self-regulating ones, that is, those that either replicate or employ without undue disturbance the normal processes of the ecosphere, and are therefore likely to be sustainable over very long periods of time; (5) the invention, promotion and application of alternative technologies which are energy and materials conservative, and which because they are designed for relatively 'closed' economic communities are likely to disrupt ecological processes only minimally (for example, intermediate technology); (6) decentralisation of policy and economy at all levels, and the formation of communities small enough to be reasonably self-regulating and self-supporting; and (7) education for such communities.[4]

The authors acknowledged their audacity in handling both the analysis of the breakdown of ecosystems and social systems and their disruption and were not suprised when they were criticised by specialists in both areas. They realised that they must appear politically naïve, but saw the necessity to combine in one analysis and prescription all of the elements of the problem. They believed it to be urgent (correctly termed, survival) material and saw how, unless the bulk of the people saw the necessity for its often sacrificial recommendations, they would be described with justice as impractical. They moved uneasily between national direction and the imposition by an informed, concerned elite, of perhaps unpopular measures, and the desirability of decentralisation and of small self-sustaining and self-regulating communities. To almost any specialist commentator looking at particular isolated proposals the document seemed unrealistic but it did form the basis of conference discussions, and educational programmes for higher education students.

Both *A Blueprint for Survival* and *The Limits to Growth* set out clearly that there were limits and that we were much closer to them than most people realised. Growth in both publications meant more than simple economic growth (in a *per capita* sense) and included population growth, but the particular aspect of economic growth was one area in which the debate was continued with particular vigour with those in opposition to the notion of a steady state of zero growth economy obviously needing to dispute the existence of limits of any kind. Since 1972 the debate in Britain has so far centred around Professor Wilfred Beckerman, Anthony Crosland and Jeremy Bray who argue that growth can and must continue, and E. J. Mishan, H. V. Hodson, Alan Coddington and Peter Victor, who argue the opposite case. John Maddox, the former editor of *Nature*, who had joined in the overall debate in 1972 with the publication of *The Doomsday Syndrome*, supported the case for growth but added little to the basic economic arguments of the three already mentioned. Dr E. F. Schumacher—leading exponent of the development of technologies appropriate to our real needs—also voiced his rejection of the no-growth argument; but, because he is aware of and concerned about the damage associated with technological and economic growth, one can hardly place him with the growth lobby. It is difficult to picture what a zero-growth economy might be like and to see what satisfactions and challenges it would hold. It is clear that its advocates see it as a dynamic condition rather than stagnation.

The summer of 1972 saw an impressive extension to the debate: the United Nations Conference on The Human Environment took place in Stockholm with a blaze of publicity, as the high point of what the U.N. had designated World Environment Year. The debate had now reached its most impressive international forum. It was accompanied by a large number of publications and reports but these represented a gathering together of well-established data and lines of argument for presentation to politicians rather than new material. The U.N. secretariat commissioned René Dubos to collect as much relevant material as possible and to distribute an unofficial report to all the participants; this was subsequently presented in book form as *Only One Earth* (1972) by René Dubos and Barbara Ward. Seventy eminent consultants prepared material for this book; their strongly expresssed views were often contradictory and the authors blended the contributory material together so that it does not bear the stylistic symptoms of having been written by a committee. It is seen by some as much too pessimistic and by others as too innocuous and lacking in that dynamic impact which they believe the case which it presents deserves and needs. The Stockholm conference was followed by further U.N. conferences on the use of the sea, on food and agriculture, and on population. To most observers the most remarkable feature of these conferences is the enormous gap between the powerful speeches of the national delegates and the apparent inactivity of governments particularly as far as taking unilateral action where agreement cannot be reached on concerted action by all the nations involved is concerned. The debate at least is now an established fact of our world scene and the quantity of published material now available is almost overwhelming.

4.2 Positions in the Debate

The foregoing account of the development of the environment debate reflects the continuing importance of contributors from the world of science. Others—philosophers, sociologists, economists—have also written on environmental themes. This section draws on this wider range of contributions in an attempt to define the different stances which have been adopted.

Chapter 5 makes the point that environmental problems are outcomes of human actions. Within the framework of institutions, values and laws that define a society, man responds to the choices which face him. It is these decisions on the allocation of time, effort and resources that determine industrial methods and the level and composition of output. As chapter 6 explains, these choices shape our environment in the broadest sense of the term—the rate of resource depletion; the nature and extent of pollution; the distribution of population; the range of manufactured objects which surround us. Whether they make this point or not, it is not suprising, therefore, that those who have written about environmental problems should seek their causes in the character of society.

Individual Moral Failure. Edwin Dolan identifies a position which he calls 'environmental evangelism' and which sees individual moral failure as the source of the problem.

> At the root of the environmental crisis the ecological evangelists see the sins of ignorance, indifference and greed. ... The content of the writing ... reflects their analysis of individual moral failure as the source of the problem. ... Their first principle is that ignorance is no excuse, that it is immoral to take any action, to produce or consume any product without knowing what effects that act of production or consumption will have on the environment.
>
> They realise that ignorance is not always the problem, and that, in fact, some people exist who are so insensitive to the beauty of nature that they would knowingly prefer their cheap lawn furniture to a noble stand of redwoods; would prefer a leopard lined with silk and flaunted on Fifth Avenue in New York City to one creeping through the jungles blending with light and shadow; that they would prefer the convenience of cheap newsprint to the beauty of clear running streams and rivers; would prefer

the thrill of gunning their four hundred horsepower coffin from stoplight to stoplight to having the tonic of clean air to breathe.

... After having disposed of the ignorant and indifferent, the ecological evangelists turn their energies to the last and most recalcitrant group, those avaricious and malevolent souls who would willingly rape the wilderness to turn a profit for themselves.[5]

Lack of a Moral Belief System. Other writers also focus on values and morals, but at the level of society as a whole. Thus Daniel Bell sees the deepest challenge to the survival of American society arising from its lack of a rooted moral belief system.[6] The point is amplified by William A. Weisskopf:

This is the situation we find ourselves in now. The main institutions and activities of Western Society, technologically oriented science, technology and the economy exist for their own sake. They have no justification on any superordinated, all embracing value system. Their only *raison d'être* is their own self-perpetuation. The main reason why people tolerate them is that their individual existence is enmeshed in these institutions and their livelihood and security depend on them. It is hardly surprising that under these conditions the more sensitive groups of modern youth feel deeply anomic, frustrated and alienated and that they try to escape the repression of a system which has lost all basis in any faith, belief, central world outlook or value system.[7]

S. H. Nasr makes a similar point when he says that 'Although science is legitimate in itself, the role and function of science and its application have become illegitimate, and even dangerous, because of the lack of a higher form of knowledge into which science can be integrated and the destruction of the sacred and spiritual value of nature.'[8] This absence of a sense of harmony with nature in Western society is, says John Black, a characteristic of the Western world view having its origin in Hebrew theology and its Christian interpretation. The role of man was to exploit nature to his own advantage, and[6] ... if Christianity will be shown in the end to have failed ... [it will be] because it encouraged man to set himself apart from the rest of nature, or at the very least because it failed to discourage him from doing this.'[9]

The Christian historian Lynn White (Jr) endorses this analysis and goes on to express his doubts about the capacity of our present science and technology to provide an answer, because they

are so tinctured with the orthodox Christian arrogance towards nature that no solution for our present ecological crisis can be expected from them alone. Since the roots of our trouble are so largely religious, the remedy must also be essentially religious, whether we call it that or not.[10]

He presents an alternative Christian view that he believes was expressed most simply by St Francis of Assisi (surely the patron saint of ecologists) who viewed the created world as God's world in which man and nature were intended to glorify Him in harmony. Some modern Christians, sensitive to the ecological impact of their individual and communal activities, look for a guide that is more fundamental and reliable than the best attempts at ecological auditing. They are rediscovering this guide in the traditional teachings of the Christian Church as they apply them, not only to man's

---

Box 4–2   *Alternative Technology*

Between those who welcome advances in technology as essential ingredients to our problems, and those who regret technology, seeing it as a root cause, stand the advocates of alternative technologies. They use a variety of terms—intermediate, appropriate or low impact technology—and generally direct their attention to developing countries and rural situations. The Intermediate Technology Development Group, founded and led by Dr E. F. Schumacher, is an example of a well-organised productive association of experts who have now a wide range of successful products to their credit. Schumacher argues that advanced technology is capital rather than labour intensive and so inappropriate to most situations in developing countries where the need is for simple rugged equipment which can be used and maintained on local resources, and is readily available. One of these working groups of I.T.D.G. developed a machine for the Zambian government which produced egg packing trays that were more robust yet simpler and which could be made from local waste paper. The smallest existing machine at that time (1969) was expensive (£170 000) and produced more trays in a month than Zambia needed in a year. The new machine cost £8500 and produced an appropriate number of trays, but could also be used to make other packaging. In addition the new trays were fully interlocking and so did not need to be packed into boxes for transportation and yet gave greater protection. At a simpler level many of the group designs are based on bicycle parts and can be made up by local craftsmen, so avoiding expenditure outside the country, and giving more people work.

relationship to his fellow men, but to his natural surroundings as well. St Paul's list of the fruits of the Holy Spirit includes patience, self-control and gentleness, and Schumacher points out the relevance of the cardinal virtues of tradition—prudence, justice, fortitude and, most significantly, moderation.[11] (See also box 4–2). John Taylor explores the theology of non-violent, harmonious and sacrificial living and gives us the modern text, *Enough is Enough*.[12]

The Anti-science Movement. Of course, it could be that the religious beliefs referred to by John Black are not the origin of man's attitude to the world around him. Both may be reflections of man's basic nature. For example, it has been suggested that the interest of primitive societies in magic reveals a basic urge in man to control his environment.[13] From this point it is only a short step to those who see science and technology at the root of environmental problems. Analysing the anti-science movement, Edward Shils summarises as follows the bill of particulars against science and technology.

> Science and its offspring, technology, are charged with squandering and despoiling an endowment which should properly serve mankind forever. They are charged with the exhaustion of stocks of natural resources needed for future generations, the defacement of the natural landscape, the disruption of the natural balance of the ecological system. Science is charged with an as yet not realised intention to tamper with man's genetic endowment, to breed creatures, perhaps human beings, of some scientist's own devising. Half-heartedly, without as yet firm conviction, it is intimated that science is at fault in the 'population explosion'. Hideous preparations for chemical and biological warfare are held against science. And underlying all of these grievances is the deep apprehension about nuclear weapons and the dangers of radiation from the civilian uses of nuclear energy. These are some of the allegedly actual and prospective consequences which are attributed to science.[14]

Many examples can be given of variations on this theme. Interviewed by Ann Chisholm, Barry Commoner ascribed the origin of the environmental predicament to a scientific time-bomb planted 75 years ago.[15] The explosion of scientific knowledge between 1900 and 1945, and the subsequent period of application, are the prime causes of the problem in his view. Even so, some would remain optimistic, like Harvey Brooks who writes: 'With respect to the great modern problems—what I call the four P's—of population, pollution, peace and poverty, it may be that articulating these is the most important part of the problem, that once these needs are formulated in the right way, the technological solutions will become obvious, or will fall into place.'[16] Eugene S. Schwartz, however, represents the opposite opinion when he says that the crisis threatening human civilisation is inherent in science and technology and is not amenable to rectification by more science and technology.[17]

The Pace of Change. A slightly different view is that the problem is not presented just by the products of science and technology but also by the pace of change. A point commonly made by environmentalists is that there is a need for caution, adaptation time, a breathing space in which to understand the ecological impact of technological developments. François Hetman observes:

> In a sense, mankind seems just to be overriden by the tremendous development of scientific knowledge and technology. This leads to a theory that while science and technology have registered an exponential advance since the scientific and industrial revolution, ethical progress—among all people including scientists and technologists—has been only very slow, irregular and hardly measurable at all. According to this view, discontent with technology can be ascribed to a bio-psychological gap between the impact of technology on man on one side and the lack of ethical understanding and assimilation of technology on the other. This gap explains why technology has always had its own way. Whenever technology has become available, the urge to use it has been overwhelming and easily justified by economic profits or political and military advantages. Little or no concern has been shown for inconvenience or side-effects of new technological developments.[18]

Technological Values. But is the need to keep up with technology really the problem, or is the real problem, as Lewis Mumford believed, that we have got to the position where we think in such terms? The view being expressed here is that it is not just the products of science and technology which threaten. An unbalanced development is taking place of that side of man's personality which parallels the values implicit in science and technology—namely efficiency, organisation, precision, objectivity.[19] Believing that the natural desire to control the natural environment is being extended to man himself, Mumford fears that man is becoming a passive, machine-conditioned animal to be fed into the machine or into depersonalised collective organisations. Dubos shares this concern that it is man's adaptation, the quality of his life, rather than his extinction, which is the more immediate danger.[20]

Defective Political–Economic Systems. A more prosaic view than interpretations which attribute environmental problems to individual moral failure, or the blind march of science and technology, is that which looks to defects in everyday political–economic processes. Thus Kenneth E. Boulding calls for a questioning of the assumptions underlying conventional economic systems. Essentially he describes them as 'mines-to-dumps' or 'cowboy' economies. The conversion of resources into goods and services, which is seen as the fundamental economic process, ignores any limits to resources and any unwanted by-products of production and consumption. The term 'cowboy economy' describes the feeling of living on a limitless plain, where rubbish can be 'thrown over the fence' and where new pastures are always to be found over the horizon.[21] Translated into the reality of the present day, his point is that our economic systems fail to take proper account of information about resources and wastes.

The Costs of Economic Growth. A very similar position is adopted by those who emphasise the environmental consequences of economic growth. E. J. Mishan, for example, accepts that indices of economic growth may be rough measures of the increase in productive power, but:

> no provision is made in such indices for the *negative goods* that are also being increased; that is, for the increasing burden of disamenities in the country. Nor can they reveal certain imponderable but none the less crucial consequences associated with the indiscriminate pursuit of technological progress.[22]

Hodson goes even further in cataloguing the environmental impact of H.V. economic growth, to include the social and cultural environment.

> The diseconomics of growth comprises, then, an economic criticism to the effect that, even within the limits of what can be measured in money, the cult of growth involves unseen distortions, misuses and wastes. It also comprises a moral criticism, to the effect that the growth cult runs counter to moral and spiritual values. It is inherently selfish; it neglects man's duty to his neighbours: it debases other parts of man's being than the economic. It encourages a mechanistic outlook on the world—asking what can it be made to do, rather than how does it naturally behave—and this leads to a mechanistic view of man himself, as a machine for producing and consuming. It compounds the sins of envy and greed: it goads men and women on to over-competitiveness and material emulation, and this warps their natural personalities, often to the extent of physiological strain or mental illness.[23]

As these extracts imply, some writers attribute environmental problems not only to specific defects within, but also to the whole character of, certain political–economic systems. So, speaking of the modern market economy, Weisskopf says: 'It could be said that this society has turned the slogan "the ends justify the means" to read that (rational) means justify (bad and irrational) ends.'[24] In other words, everything that science and technology make possible is permitted and justified because it can be done efficiently and profitably. Similarly, Dolan reports a view typically held by those he calls the environmental radicals: '. . . as long as society organises production around the incentive to convert man's energies and nature's resources into profit, no planned, equable, ecologically balanced system of production can ever exist.'[25]

At this point it must be admitted that, of course, the positions defined in this brief review represent differences of emphasis more than entirely different categories. In several cases there is clear overlap between the examples given, and there could even be dispute about the categories into which they should fall. However, attempting to define various positions on the general causes of environmental problems is a useful exercise. Also, as the next section will show, they can be used as a framework for an analysis of possible approaches to environmental policy.

4.3   Approaches to the Solution of Environmental Problems

Broad approaches to the solution of environmental problems are implied by the categories used in the last section to describe positions adopted in the environment debate. These were: individual morals; a society's prevailing cultural and value system; scientific and technological development; defects in the functioning or structure of political-economic systems. Proposals to solve environmental problems are thus likely to include one or more of the following elements:

(a) changes in the attitude and behaviour of individuals;
(b) a cultural revolution;
(c) the assessment and control of science and technology;
(d) specific reforms in society's decision-making processes;
(e) radical political–economic changes.

Individual behaviour might be regarded as supplying the fundamental building blocks of social processes. In that environmental problems are the outcomes of social processes, it can be understood why some writers, when suggesting solutions, emphasise the role of the individual. 'The only solution to the environmental crisis', Ivan D. Illich believes, 'is the shared insight of people that they would be happier if they could work together and care for each other.'[26] 'In the end', Dennis Pirages and Paul Ehrlich say, 'each person must be made to feel responsible for the present and future welfare of all mankind.'[27] This last remark seems to suggest that the changes in individual attitudes will be engineered at the level of society as a whole. It is this sort of cultural change which is described by Lynton K. Caldwell: a new ethic is needed, a theology of the Earth; a powerful political ideology could emerge from a view of man in nature arising from a convergence of science and religion.[28]

But which comes first: a change in the individual, or change in the social system which surrounds him? The same dilemma presents itself when considering the environmental impact of science and technology. Some see the need to overcome scientific philistinism by making scientists and technologists more sensitive to human rights and values. But what incentive, responsibility or authority have they to take into account broader social perspectives? It is likely that the behaviour of scientists and technologists will change only if the framework within which they operate is changed. New institutions and procedures would be needed for the control and management of science and technology. An Office of Technology Assessment, for example, might be established to survey the actual and potential effects of technological developments as a basis for community policy. (See box 6–5 for an outline of the functions of such a body.)

The analysis underlying this approach is that important information has been left out of the calculations which guide scientific and technological development. The suggested solution was to generate the additionally relevant data and feed them into the decision-making process. This is likely to involve new constraints on the freedom of action of those engaged in research and development, the creation of new institutions, and probably an extended role for the government. The same sort of analysis can be applied to other environmental issues like resource depletion and pollution, and parallel solutions can be proposed.

It is the task of a society's political–economic system to guide the allocation of resources. In Western societies a key role is played by the market, in which consumer preferences and the relative scarcities of resources are reflected in market prices. In principle, guided by these signals, the independent actions of millions of individuals and organisations should result in a rational use of resources. Yet pollution, for example, still occurs. How can it be that actions which are individually rational add up to something which is collectively irrational? (This issue forms the main theme of chapter 6.) An obvious possibility is that some relevant information has been left out of the decision-making processes. Perhaps some costs, or some benefits, have been ignored—such as the costs to the community of polluted air. The industrialist, or motorist, has not been presented with a bill for his use of the atmosphere. So, his decisions about production methods, levels and locations, or in the case of the motorist, where he lives, how much he travels and by what means, have been distorted.

One response to this sort of problem is to refine the data on which decisions are based. Dolan outlines

> a constructive strategy for coping with the environmental crisis. The fundamental principle on which this strategy is built may be expressed in a simple slogan—There Ain't No such Thing As a Free Lunch—the 'TANSTAAFL principle' for short. The TANSTAAFL principle is closely related to the fundamental theorem of ecological economics, that everything depends on everthing else. Everything worth while has a cost. Whenever you think you are getting something for nothing, look again—someone, somewhere, somehow is paying for it. Behind every free lunch there is a hidden cost to be accounted for.

> The task of ecological economics is to figure out how to restructure the economic system so that these hidden costs will be brought into the open, with the ultimate aim that no one who benefits from the use of the environment will be able to escape paying in full.[29]

Such hidden costs might be made visible in the form of a pollution tax. The expense incurred by an industry in conforming with a regulation, for example, on the quality of waste discharged into a river, would have a similar effect. What previously appeared free would now have a price tag.

The solutions suggested by those who place economic growth at the centre of the environment issue are necessarily so radical in their implications that they are arguably more than just reforms. A no-growth or steady-state economy (discussed further in chapter 7) implies a fundamental change in attitudes. This is postulated by Pirages and Ehrlich: 'The task for the future is clearly one of restructuring economic norms, values and habits to move away from an economic model that encourages increased consumption of resources towards a model that limits growth in such consumption. We must also ensure that economic growth is carefully monitored and channelled in desired directions.'[30] What would a no-growth economy entail—population control? resource rationing? a more labour intensive technology, a smaller workforce or much reduced hours of work? some criteria for the distribution of incomes other than those based on reward for productivity? an increased share of resources for one purpose or group in society only at the expense of another?

Such an ecologically managed society fails to satisfy some writers because of the bureaucratic totalitarianism which it threatens. Dubos, for example, argues that such a 'highly structured and unified environment may be desirable for the sake of order, efficiency and peace,' but he stresses that 'diversity in the social environment should be made one of the tests of true functionalism.'[31] Illich makes a similar point that in his *convivial society* people should define their own images of the future and not leave it to experts and politicians. Science and technology should be used to provide decentralised opportunities for efficient, creative and self-directed work.[32]

Radical proposals like these for restructuring political–economic systems are naturally depicted only in outline by their authors. They are interesting and challenging because they seem to imply, paradoxically, an unprecedented degree of control at the level of, and between, societies—over-population and global resource depletion and allocation, for example—and also of personal freedom—of access by individuals and small groups to their own sources of energy, and to direct their own work.

Summary

The environment debate as we know it today is a phenomenon of the period since the Second World War, although related writings on particular issues like population and conservation can be found from the nineteenth century onwards. This is not to say, of course, that environmental damage by man is itself a recent phenomenon.

In section 4.1 we have outlined the development of the modern debate from Ehrlich and Commoner in the 1960s, through *The Limits to Growth, A Blueprint for Survival* and the Stockholm Conference. Widening the focus beyond these major contributions, section 4.2 attempts to define various positions adopted in the environment debate. The categories suggested are: individual moral failure; the anti-science position; the pace of scientific and technological change; the spread of technological values; defective political–economic systems; the cost of economic growth.

In section 4.3 this categorisation of positions in the debate is used as the basis for an analysis of the possible approaches to the solution of environmental problems. We have suggested that proposed solutions are likely to contain one or more of the following elements: changes in the attitudes and behaviour of individuals; a cultural revolution; the assessment and control of science and technology; specific reforms in society's decision-making processes; radical political–economic changes.

The remaining chapters of the book offer a more comprehensive analysis of the possible causes of, and solutions to, environmental problems by concentrating on the framework of society within which they occur.

CHAPTER 5

The Politics and Economics of the Environment

Everything seems to point to the complexity of the environment problem. As individuals we are aware of the blurred edges of the issue in our own minds. Starting with a particular problem, our attention spreads outwards like the ripples on a pond. This experience reflects the nature of the ecosystem and is reinforced by any acquaintance with the principles of ecology. Alternatively, even if one is unaware of the maze of interrelationships, the complexity appears in the form of a long list of problems—population growth; air pollution; extinction of plant and animal species; soil degradation; resource depletion; water pollution, and so on. Is there a common element, a central issue? The prospect of finding one is attractive, but is it realistic? The answer is probably no if the search takes the form of a detailed examination of each issue to find a technical or scientific common denominator. However, looking not at the internal composition of problems, but at their external effects, the simple point emerges that they all affect man. Put another way, the problems are defined by man.

Although simple, even obvious, the point that all environmental problems are human problems, in the sense that they are defined by people in terms of their effects on themselves, is not without significance. It means that environmental problems are not acts of God or plagues visited on us by nature. Generally, the situations arise out of our activities, and it is we who designate them problems. Given that we are rational, there might appear to be a contradiction here in that we are creating problems for ourselves. One explanation could be that the problems are unintended and unexpected side-effects. Alternatively, we are assuming that the decisions to undertake activities and then apply the problem label to their outcomes were taken by the same body. Is it possible that different decision-makers were involved? Whichever explanation accounts for the generation of environmental problems, it is necessary to examine the way in which decisions affecting the environment are made.

5.1 Choice in the Use of Resources

Progressing logically, if an individual identifies a problem then something must have been done which he does not like, something which clashes with his preferences. For anything to have been done effort and resources must have been expended, so to identify an environmental problem is in essence to say that resources have been used in a way which you do not favour. For example, the spread of urban development into the green belt means that land has been used for commercial and residential rather than agricultural or recreational purposes. The rapid depletion of a mineral resource versus its conservation represents a conflict about the rate of utilisation of a resource over time. The decision to discharge effluent into a river represents a choice between different uses of the river resource, and also between the use of the river and the resources involved in alternative methods of disposal. The issue of population growth can be expressed in similar terms. To support population control is to favour the possibility of concentrating resources on a smaller number of people now or spreading out their use over a longer period of time (allocating them to more distant future generations) than could be done given the present population size and growth rate.

In effect, the social organism (the interacting collection of individuals and institutions) has decided on the allocation of resources. In that environmental problems are the outcome of resource use, then society has determined the creation of the problems. This is to say that environmental problems are essentially man made. They are things we do to each other or to future generations. This simple fact is extremely important. It represents a foundation on which a widespread understanding can be based, and provides a starting point for analysis leading to action. For example, we might ask what decisions affect the environment? who takes them? and what are the economic, political and social forces behind the decisions?

In any society, making decisions about the allocation of resources is the central task of the political–economic system. A textbook on economics, for example, is likely to start with the proposition that resources are scarce in relation to the output desired from them. The need to maximise output is taken for granted and certain choices must then be made. Different combinations of goods could be made with given resources: what relative priorities should be assigned to them? For any given product, different materials and production methods could be used. Which would be the most economical? This concern with resource allocation does not mean that economic theory has always focused on environmental issues as defined today. Its starting point has been the desire for an output of goods and services for distribution among the people and it has tended to emphasise the need for growth in order to raise output per head. But in terms of contemporary environmental concerns, this suggests the long-term incompatibility of continuous economic growth and a limted stock of resources. In conventional economics the resource problem has beeen presented as one of limited availability over a given accounting period. In other words, attention has been on the flow of resources becoming available in a given period of time, rather than the stock. This is not to say that economic thinking has been intrisically inadequate. Its focus is logical given that the period of time it has been concerned with is short. This is near-sightedness reflects the weight we all give to the fairly immediate future, and it is not inconsistent with the concern now being expressed about the total stock of resources, which arises when the long term is thought to be merging with the short.

So, long before the environment issue took on its present-day form, social sciences, such as economics, were studying the processes of resource allocation: how a society ordered its priorities and chose its output pattern, and how the decisions were made on production systems, methods and resource combinations. The connection between these everyday political–economic processes and current environmental concerns can be seen by looking more closely at the decisions involved and their practical implications.

The Level and Composition of Output. The higher the level of industrial or agricultural output the greater the demand on resources and the higher the level of pollution potential. Given the methods and efficiency of production this is true because output is a function of resource inputs, and so is the generation of unwanted side-effects. The same basic relationship is presumed to exist, given the size of the population, between the level of output and the average material living standard in a country. So, the simple point to be made is that decisions about the level of economic output will affect the environment both in terms of benefits (output of goods and services) and costs (pollution, resource depletion). But, if economic growth creates both costs and benefits, what is the relationship between them? Are the costs significant or incidental? Do they bear a constant relationship to benefits or do they grow faster than benefits after a point? This question of the link between economic growth and human welfare has become a major strand in the environment debate.

Those who question the benefits of economic growth also make the point that our welfare is influenced not only by how much, but also by what is produced. At the most general level, take, for example, the division of output between consumer goods and capital goods, (like roads, machinery, buildings), or between private and public sector goods in a mixed economy. An increase in the proportion of resources going into capital goods would defer immediate consumer benefits, and in the case of a shift of resources to public rather than private goods, personal expenditure on clothes and consumer durables would be exchanged for public expenditure on education and health services. Whether one mix of private and public goods was considered to produce a higher level of human welfare than another would depend on the values held. There could, however, be consequences for the environmental impact of production in terms of resource depletion and pollution if public goods were characteristically different from private in their resource demands, for example, service activities like education with low material inputs relative to labour.

The same points about different contributions to human welfare and impacts on the environment could be made about the particular products which make up either public or private spending—nuclear weapons, the motor car, Concorde, D.D.T. and so on. The fundamental point is that environmental problems can be set within the framework of everyday political–economic decision-making, and need not be treated as a separate issue.

Production Systems and Methods. How can the production systems of contemporary Western societies be characterised—large-scale industrial production, heavily capitalised and with a severe division of labour, production being organised mainly in privately owned, professionally managed and publicly regulated corporations? Like any generalisation, this is not an exact description of reality, but it serves to illustrate a fundamental point. There are various ways of organising production—subsistence farming, small-scale cottage industry, state ownership and control, production communes—of which this is only one. It should be apparent, however, that the system of production has a pervasive influence on the human environment.

By comparison with the others, the modern industrial system has a capacity to produce a much larger and more varied output. Technology, mechanisation and accompanying developments in social organisation have enabled it to tap the Earth's energy reserves. Of course, it has a parallel capacity to use up resources and create wastes. It might also be argued that it exacts a direct human price by creating highly urbanised and rapidly changing societies.

The character of social relations is also influenced by the fact that in industrial societies people live and work together in large numbers; most jobs are highly specialised; ownership and control of industrial activities are concentrated; workers do not own their output and often do not control the pace at which they work.

Within the framework of the industrial system, routine decisions also influence the human and physical environment—the choice of materials, production and waste-disposal methods; the scale and location of operations; the direction of research and development.

Who Gets What? In response to the limited availability of resources, having determined what to produce and how to produce it, the remaining general task of political–economic systems is to share out the proceeds. Conventionally this has been seen as a problem of distributing goods and services. Fundamentally a person's access to goods and services, and his freedom of choice about where and how he spends his time, bears some relation to the distribution of income and wealth. The same can be said of environmental *bads*, which might be seen as the absence of goods such as clean air and beautiful surroundings (see box 5–1).

Box 5–1   Environmental Fairness

'Whenever social costs are shifted on to the economically and politically weaker sections of society without compensation, a redistribution of the costs of production, hence of real income, is involved. During the first Industrial Revolution it was undoubtedly the industrial proletariat that carried the burden of social costs, in low wages, long hours, high accident rates, social insecurity, etc. The present environmental dangers threaten all sections of the population but in unequal measure. The higher and middle income groups are able to evade the worst impact of pollution, noise and traffic chaos by moving to suburbs in the green belt areas, or to smaller towns, or by the installation of air conditioning, etc. The poorer sections and the ghetto population have no means of evading the unhealthy working and living conditions, and are more exposed to noise, traffic chaos and pollution, with far less possibilities for recreation. For example, the concentration of toxic substances such as carbon monoxide and oxides of sulphur is ten times higher in cities than in the country. In the city centres of the United States, which are more heavily populated by blacks and other minority groups, these toxic levels are much higher than in the suburbs'—K. W. Kapp[1]

The point made here, and throughout this section, is that environmental problems do not exist in the abstract. They are defined in terms of their impact on humanity, so can one think of their impact on man independently of their impact on men? In order to measure the extent of the problems or assess the success of policy, reference must be made to how environmental bads are distributed between people. In this case, as with the environmental implications of choices on the level, composition and methods of production, it is important to realise that environmental issues are not separate. They can be seen as an integral part of ongoing political–economic processes. Analysis of these processes throws light on the origin of environmental problems and on possible reactions to them.

5.2   Decision-making Systems

The last section outlined the areas where, in principle, choice must be exercised by any society. In practice these choices must be based on some criteria which permit a distinction to be made between less and more desirable alternatives. Also, preferences must be made effective. Choices having been made on the best use of resources, this information must then be transmitted to those who have control over resources. The aim of this section is to introduce those fundamental issues which are involved in society's decision-making systems. What are the origins of the preferences on which choices are based, and how are they expressed?

If decision-making is not to be a random process, it must be based on some data relating to costs and benefits. In all cases, it could be argued, these costs and benefits must be measured by reference to man, there can be no independent yardstick. Ultimately the cost of a resource is the benefit which would have been derived from the best alternative use of it—its opportunity cost. And the concept of benefit relates to man either in the sense that an action has a favourable physical effect on him, or that it conforms to his values. Thus food production is a benefit, and so is the preservation of rare animal species if man believes that he owes, to God or future generations, a duty of stewardship over the Earth. In many cases a benefit may be both concrete and abstract. The preservation of genetic diversity is such a case, in that it also has potential direct advantages to the present population. Equally, it is possible for the two to conflict as illustrated by a desire to exploit a resource—for food or energy—as well as preserve it for future generations. Similarly, with many products there is a clear possibility of conflict between the desire for more output and for less pollution.

In these circumstances, even if there was only one decision-maker acting on behalf of a society, there would be a need for a complex weighing of costs and benefits. The matter is infinitely more complicated if it is accepted that the various individuals and groups of which a society is composed have each a legitimate point of view. If each individual has a unique schedule of preferences—on the uses to which resources should be put, and what weight should be given to the interests of future generations—how are they all to be expressed, co-ordinated and converted into action?

Based on a value judgement about the primacy of the individual, one theoretical extreme is represented by the market approach. Each individual would express his personal preferences on the allocation of resources through the pattern of his own spending. Producers would react to the demands expressed in this way, organising resources to meet them and being rewarded by a surplus of sales revenue over their costs—by profits. If demand was not fully satisfied then the relative scarcity would push up prices, increasing the rate of profit and stimulating more production. If, on the other hand, supply was excessive, competition to sell would force down prices, profit margins and production. In this way, through the medium of producers seeking profit, resources would be continuously channelled into activities which matched consumer preferences. Of course the adjustment from one pattern of output to another would not be instantaneous, but the tendency of the system would be to reward those producers who were using resources for the *right* purposes, and penalise those who were not. Where the consumers of different products were in competition for a scarce resource (energy, materials, labour) it would be used by businessmen in the manufacture of that product in which it would earn the highest return. The theory of the market would assume that the satisfaction yielded by a product was reflected in the price paid for it. It would thus be argued that the market mechanism was guiding resources into those uses which would promote the greatest benefit.

In the theory of the market, similar mechanisms would determine the choice of materials and production methods. Faced in the manufacture of a particular product with a choice of materials, combinations of labour and machinery, size and location of factories, and sources of energy, the businessman would try to minimise costs. His desire for profit would spur him, and so would competition. The relatively inefficient producer would be penalised by lower profits or even losses, and would leave the business or improve. In this way scarce resources would be economised. Where different industries had a shared need for a scarce material, competition between them would force up its price. Eventually, one would find it cheaper to substitute another material, and the original material would go to the other industry, the one least able to use an alternative.

Although the market model may sound unreal as just described, it merits attention because both in theory and practice it represents a major approach to the resource allocation problem, and it frames most thinking about economic and environmental issues in Western societies. It occupies this central position for several reasons:

(1) As a theoretical system it has an attractiveness based on its simplicity and the fact that it appears to offer rational solutions to complex social problems by harnessing a presumed basic drive, private self-interest, in an automatically functioning mechanism. It would seem that conflicts are resolved as if by a 'hidden hand'.

(2) It is thought by many to match certain basic political values such as freedom, individualism, democracy.

(3) It incorporates some fundamental ideas about rationality in resource use.

(4) It is embodied in an existing set of social and economic relationships which have inertia.

Of course, it must be repeated that this model of the market mechanism represents a theoretical extreme. In practice things are not quite so simple. In the realm of social and political controversy there is a long history of criticism of the consequences of untrammelled market forces, expressed in terms such as inequality, monopoly, privilege, exploration, etc. Also, even apart from such criticisms, the very presence of serious environmental problems such as pollution and resource depletion suggest that the market has its faults. Section 5.3 gives some examples of this gap between model and reality.

Political–economic systems are often depicted as forming a spectrum ranging from the totally decentralised market to the complete central control of a command economy. In one case the origin of preferences is the individual and they are expressed directly through the automatically functioning market. In the command economy, a central body would determine collective requirements. The broad decisions of a governing elite on the pattern of production would be translated into practical terms by a planning body, and transmitted to productive enterprises via a command structure. Such a bi-polar representation clarifies the differences between political systems but tends to exaggerate them. For example, the ruling elite of a command economy would probably claim to be interpreting the will of the people, perhaps through a party machine. Similarly, in the market system, business enterprises might not be entirely passive recipients of signals from consumers.

They might filter them, by responding selectively, or exert a modulating influence through advertising.

Practical systems are likely to contain elements from both the decentralised and command models. Thus Western societies use market mechanisms, but giant firms are able to exert influence, and of course state organisations may occupy a central decision-making position in areas such as military spending, education, health and welfare, fuel and transport. The purpose of this section of the book has not been to discuss political–economic systems in detail, but to illustrate essential decision-making issues and the range of approaches to them. In examining environmental problems it is essential to consider these decision-making processes. Environmental problems are not a special category, for they are the outcome of conventional political–economic processes. Thus, as later chapters will show, most responses to them take the form of proposals for reform or restructuring of the political–economic system.

5.3 Unintended Consequences and Conflicts of Interest

How is it possible that the presumably rational decisions of men can result in the creation of environmental problems? It was suggested in the introduction to this chapter that everyone's apparent interests may have been considered, but that problems arose in the form of unintended side-effects. Alternatively, perhaps some groups in society were able to express their preferences more effectively than others. In this case, outcomes desirable to one group could be termed as problems by another. This section considers these possibilities in the context of Western economies.

In principle, the market mechanism should govern the interaction between the consumers of a product and the rest of society in such a way that production takes place up to, and not beyond, the point at which the benefits derived by the consumers are equalled by the consequent costs borne by others. This balance should be reflected in the price of the product, representing both the benefits to the consumers and, in that it covers production costs, the necessary compensation to all those in any way involved in, or affected by, the production process. In practice, the apparent cost of production, on which the selling price is based, may not take into account all the true costs involved. This could happen due to ignorance of some of the costs, or the inadequacy of the market mechanism. (See box 5–2, figures 10 and 11.)

The impact of a pollutant may not be fully understood so that even if compensation were to be paid to all those concerned it would be too low. This

Box 5-2   Costs, Benefits and Hidden Costs

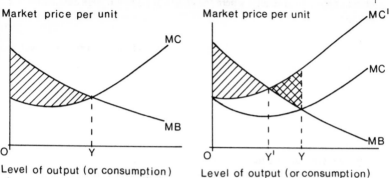

Figure 10   Costs and Benefits

Figure 11   Hidden Costs

Market Price per Unit represents the costs incurred (or benefits derived) per unit of production (or consumption) shown plotted against output level.

MB = Marginal Benefit; the benefit derived from additional consumption of a product at different levels of total consumption.

MC = Marginal Cost: the costs associated with extra production at different levels of total output (conventionally accounted).

In figure 10, at output (consumption) levels lower than OY, marginal benefit exceeds the marginal cost. Beyond OY each unit of production (consumption) will incur costs which outweigh the associated benefits. The output level OY thus defines the point at which net benefit is maximised (shaded area).

In figure 11, the vertical distance between MC and MC^1 represents additional costs which had not previously been accounted for, due either to ignorance or to obstacles to the representation in the market of those who bore them. On the basis of this true level of costs (MC^1) production (consumption) at OY is beyond the level (OY^1) at which net benefit is maximised. Thus, any reduction in output (consumption) from OY towards OY^1 will reduce the area of net costs (cross-hatched) until the point OY^1 is reached at which the area of net benefit (shaded) is maximised.

The difference between MC and MC^1 can be regarded as a hidden subsidy which results in a higher level of output (at a lower price) than would have been the case if the producers were faced by the true level of costs. This point is made in chapter 6, where the possibility is also discussed of introducing a tax to represent the missing costs.

could happen most readily where the effects of pollution were cumulative, for example:

(a) If there is ignorance of the cumulative effects of exposure over a period of years, wages for industrial workers subject to lung disease would be inadequate to compensate for conditions.

(b) The costs associated with the dumping of a toxic material might be based on the assumption of cumulative but slow harm, whereas the level of damage might accelerate in a series of steps as critical thresholds were reached.

(c) Several apparently unconnected side-effects of production, each adequately reflected in costs and prices, might interact to produce additional, unaccounted for, problems.

Apart from such problems of inadequate knowledge, the market mechanism itself may be inherently incapable of the tasks assigned to it. For example, one type of market failure is commonly associated with situations where normal property rights do not exist, or where there is some kind of ownership but there are obstacles to normal market transactions. In principle, the price of a product represents the outcome of negotiation between seller and buyer. The normal situation in a market economy is that the seller has clear property rights to the article being sold, rights which he is free to sell, or not, at a price he has set or negotiated. The air, a river, or a stock of fish in the sea have the character of a common-property resource, a free or collective good. Such resources tend to be used as if their use was costless because that is how they appear to the individual user. Their use does not appear in accounted cost and does not figure in pricing decisions. Even where some sort of right does exist, such as that of a householder to reasonable absence of traffic noise, or the right of people to enjoy a river, the practical obstacles to normal market transactions may be enormous. Consider, for example, the difficulties of arranging for the individual compensation of householders by passing motorists, or of Sunday strollers by up-stream polluters.

The exploitation of a common-property resource like a fish stock illustrates the case where individually rational actions, in response to the guidance of the market, result in some undesirable consequences for people collectively. It is equally possible for an activity to generate external benefits. It is commonly argued, for example, that there are social benefits arising from educational spending. In addition to the benefits gained by the individual being educated, there are spillover benefits which accrue to the community at large. A similar case might be made for public transport, when compared with the private car, in terms of reduced congestion, fuel efficiency, etc.

If, for whatever reason, market decisions do not take into account all relevant information on costs and benefits, what will be the result? In a business with a waste-disposal problem, guided by the apparently zero price of the air or waterways as immediate sinks, companies would favour their use over alternative forms of treatment involving direct costs. In effect, some of the costs of waste disposal would have been imposed on others rather than borne by the companies and their customers, hence the terms *external costs* or *spillovers*. This represents a departure from the market ideal in that some of those who incur costs have not been compensated. This windfall from the companies' viewpoint will be reflected in prices or profit margins and more of the product will be produced, more resources will be devoted to it, than would otherwise have been the case. (See box 5–2, figure 11.) Another example of distortions in resource allocation resulting from false signals from the market-guidance system is provided by rail transport. If the prices of rail services do not reflect the true social benefits of this mode of transport, and the prices of road transport do not reflect the associated social costs (loss of amenity, noise, accidents, air pollution), then individual choices will be weighted towards road transport.

In effect, these unintended consequences result from situations where, due to a lack of knowledge or the inadequacy of the market mechanism, some interests have not been fully represented—for example, the users of a river, or the community as a whole. But what is meant by 'all interests being fully represented'? In the market model each pound or dollar has equal power to command resources, and each person is assumed to spend money in a way which optimises the benefit he derives from it. Pure market theory would take this to mean that total resources are used to best advantage, that there is no misallocation. The implications of this can be illustrated by assuming a very unequal distribution of incomes. The bulk of resources would be allocated in response to the spending of the super-rich minority. This disposition of resources would probably not match the preferences of the very poor majority which, due to their lack of purchasing power, would not be expressed effectively in the market. Thus, in this extreme case, even if functioning perfectly, the market mechanism could result in situations which a majority of the population would describe as problems.

This raises the question of whether a redistribution of incomes, by permitting the majority to express their preferences more effectively, could increase the sum of benefits derived from a given amount of resources. Economists have long been aware of these welfare issues, but have found it impossible, within the framework of accepted scientific method, to show that one distribution of income in a society yields a greater sum of benefit than another. Does the extra satisfaction of the poor, derived from money taken from the rich, more than balance the reduced satisfaction of the rich? However, despite these difficulties, if those concerned about environmental problems are not going to ignore the issue of the distribution of environmental goods and evils between people, they must give some attention to the distribution of income and wealth. Even if they did not want to redistribute the stock of environmental goods and evils (by redistributing income and wealth) in such a way as to raise aggregate welfare, the conventional equity issue can still not be avoided. For example, if an environmental improvement programme is to be financed from public funds then, even leaving aside the distribution of consequent benefits among different sections of the community, the method of financing the scheme would itself raise the fairness issue. The different taxes which might be used to raise the money would have different effects—taxes on expenditure, for example, tend to weigh more heavily on those with low incomes. Similarly, the need to select projects within such a programme would also focus attention on the distribution of environmental bads. Would it be possible, or desirable, to make such a choice without reference to the relative numbers of people affected? The point being made here is that the market process does not miraculously resolve these complicated questions of resource allocation, it only appears to offer an optimum solution if they are ignored.

In present-day market economies there are also other reasons, in addition to the distribution of incomes, why some interests might express themselves more effectively than others:

(a) The trend towards concentration in business has increased the power of corporations in their dealings with consumers and, particularly in the case of multinational companies, with governments too.

(b) Within nations, as decisions on resource allocation are increasingly taken outside the framework of the market, the ability of different groups to influence political processes becomes more important.

(c) In the international context, the massive differences in economic strength between the typical developed and less-developed country influences the allocation of resources between nations.

In classical market theory, many firms competed in every product market, each accepting the prevailing price and responding to consumer demand. In practice this competitive process, together with the demands of technology and other forces such as the desire for control and security, has led to a concentration of industrial power. In the United Kingdom and in Germany, the 100 largest manufacturing firms control 50 per cent of sales, and in the United States of America 200 corporations account for 60 per cent of assets in manufacturing industry.[2] The greater the degree of monopoly power, the greater the ability of firms at least to respond only selectively to consumer demand, or even to guide it into channels which match their production plans. This situation is particularly likely to arise where products have been introduced following a long period of research and development and a massive initial investment. Of course it is not only the pattern of production which may be affected. In Western societies, because of the central role of industrial and commercial enterprise in the allocation of resources, the concentration of industrial power is likely to influence all the fundamental political–economic issues, mentioned earlier, which affect the human environment. Those in whose hands decision-making powers are concentrated will make judgements which tend to favour their own interests on issues such as the rate of economic growth, resource depletion and pollution; the pace and direction of technological development; and the choice of production methods and industrial location.

Although the power of individual consumers may be less than market theory has supposed, parallel with the process of concentration in industry there has been a growth of trade union and consumer organisations and also of central government. John Kenneth Galbraith in his *American Capitalism* (1952) charted what he believed to be the collapse of the market model and the rise of 'managed capitalism' together with the 'countervailing power' of unions, consumers and governments.[3] In his last works,[4] however, he sees the state not as a counterweight but rather as an extension of the planning apparatus of the corporations—managing the level of aggregate consumer demand; providing funds for research, development and investment; acting as a major purchaser of industrial output. Where public regulatory bodies have been established, says Galbraith, they have become the servants of the 'technostructure'. In addition, as Charles K. Wiber argues, the increasingly multinational character of modern business makes it even more difficult to regulate.[5] The state does not have sufficient information, and legal and political institutions are inadequate to control the operations of multinational firms. The current interest in consumer organisations, industrial democracy and participatory politics reflects an awareness that this centralisation of decision-making power undermines what was presumed to be the automatic democracy of the free market. In consequence, any study of the real dynamics of decision-making must extend to the politics of power, influence and pressure, whose complexity contrasts sharply with the neat formulations of market theory.

However satisfactorily each nation resolves the conflicts of interest within it, there remains the question of the allocation of resources between nations. Environmental problems affect people, and national boundaries are from this viewpoint arbitrary. But they are nevertheless real, and the problems of

---

Box 5-3   *The Allocation of Resources between Nations*

'Today, the industrialised countries are using the bulk of the world's resources, and have a level of consumption totally incompatible with the unsatisfied needs of the two-thirds world. International justice requires that a needs-based minimum, in terms of food, housing, health and education, be assured to every person in every country. But since resources are limited it also requires a highest permissible maximum for consumption in developed countries.

For the countries of the Third World, the industrialised countries no longer constitute a hope for the future, a source of capital and technology necessary for their development. On the contrary, because of the exploitative character of their investment and technology, the activities of multi-national companies, and a world trade system which has always benefited the affluent, the industrialised countries are regarded as a main hindrance to the development of the two-thirds world. The developing countries are therefore struggling for social and economic self-reliance. The essential element of development is no longer aid, nor even trade, but liberation from prevailing economic power.

The recent oil crisis is one example of the struggle for self-reliance in the Third World. The meaning of that crisis is that the producing and developing countries working together through OPEC have taken over the oil power from the consuming and industrialised countries. By acting together the developing countries can use their raw materials as instruments of power, and can thereby force the industrialised countries to change their patterns of consumption'— World Council of Churches, Report on Science and Technology for Human Development, 1974[6]

---

resolving conflicts of interest between, parallels those within nations—interests vary widely; there can be said to be economic and political mechanisms for resolving conflicts of interest; but there are great disparities of economic and political power. Currently, most of the flow of the Earth's resources is claimed by the developed countries and one view of the future sees the sharpening of conflicting claims—for the continued growth of the already high *per capita* living standards of the minority populations of North America and Europe, or for the rapid growth of material standards for the presently fast-growing majority in Africa, Asia and South America. (See box 5–3.) Any transfer of resources to the less-developed countries depends on

---

Box 5–4   A New International Economic Order?

'The essence of the present demand for a new order lies in a realisation by the poor nations that they can negotiate a better deal at the international level through the instrument of collective bargaining. While the new trade unionism of the poor nations has yet to take a concrete and specific form, its objectives are clear: a greater equality of opportunity and participation as equals around the bargaining tables of the world. The demand for a new economic order—similar to the demand for political liberation in the 1940s and 1950s—has to be viewed, therefore, as part of an historical process rather than as a set of specific proposals at the moment.

. . . an essential element in these negotiations must be the search for a new framework of orderly resource transfers from the rich to the poor nations, which is based on some internationally accepted needs of the poor rather than on the uncertain generosity of the rich.

An element of automaticity must be built into the resource-transfer system. The world community is still in a stage of evolution where the concept of international taxation of the rich nations for the benefit of the poor nations may be regarded as unrealistic. But the concept need not be accepted in its entirety now; it can be introduced gradually over time—certain sources of international financing can be developed, such as tax on non-renewable resources; tax on international pollutants; tax on multinational corporation activites; rebates to countries of origin of taxes collected on the earnings of trained immigrants from the developing countries; taxes on or royalties from commercial activities arising out of international commons such as outer space, ocean beds, the polar region'—Mahbub ul Haq (1975).[7]

the interaction of several aspects of their relationships with the industrialised nations—their access to export markets; the ratio between their export and import prices; the inflow of official aid and private investment, together with their terms and conditions. It is in these areas that the superior power of the industrialised nations, and the multinational corporations based in them, has told; but there is a rising demand from developing countries for the establishment of a new international economic order (see box 5–4).

Summary

This chapter has focused on the everyday political–economic processes which determine the allocation of resources and the environmental problems which are the outcome of such choices. It is clear that these are not simply technical, but techno–social problems involving a complex balancing of interests. Although we have used the market mechanism to illustrate the fundamental issues, the problem is not peculiar to market economies. Even in a country where there was no private enterprise, where the market mechanism played no role (including the market represented by competitive, interest group politics), and where everyone was equally represented in the political process, some calculus of costs and benefits would still be needed. The next chapter outlines the approaches which have been suggested to refine the market mechanism which underlies Western economies, and to modify political processes through institutional innovation.

CHAPTER 6

Environmental Policy

If environmental problems are the result of resource misallocation, the spotlight of potential reform falls most obviously on the decision-making processes. Where are the decisions made and on the basis of what information? If, as in the case of the externalities described in chapter 5, relevant information has been left out of account, a ready solution suggests itself. It should be possible to influence the decision by modifying the information which is fed into the decision-making process. The term 'information' is interpreted widely here. In a market economy, taking the business enterprise as the major source of resource allocation decisions, the market prices of labour, materials and finished products clearly constitute relevant information. According to market theory these adjust automatically to reflect changing supply conditions and consumer preferences. These price signals could be modified by the introduction of taxes or subsidies. But prices represent information only in the most obvious and narrow sense. Decisions, even in an era of computers, are not taken by calculating machines. They are the outcome of the behaviour of people in the context of a legal, social, institutional and cultural framework. Modification of this framework could take various forms: appeals to save energy, new regulations on pollution, changes in the law relating to property rights, and so on.

This chapter explores issues in environmental policy through a consideration of the techniques just outlined: taxation, subsidies, and the role of law and government. Pollution abatement provides most examples, but problems of resource depletion, population growth, the control of technology, and the social environment are also included.

6.1 Policies towards Pollution

6.1.1 *A Pollution Tax*

The divergence between private and social cost has been called 'the fundamental cause of pollution of all types'.[1] Where the private accounting costs of an industrial process do not match the true costs to the community as a whole, there are said to be spillovers or external costs. In effect the externalised cost represents a hidden, unplanned subsidy to the activity involved. By comparison with the theory of optimum resource allocation outlined in box 6–1, this amounts to a distortion, with resources being used beyond the limiting point at which marginal benefit equals marginal cost (see also box 5–2). Any reduction in the output of this industry, up to the point at which marginal cost is reduced to the level of marginal benefit, would reduce costs more than benefits and therefore represent an improved allocation of resources.

Box 6–1   *Optimum Resource Allocation*

Ideally, whatever the form taken by the decision-making processes, they should take into account all the benefits and costs associated with a particular use of resources. In principle it would then be possible to achieve an optimal allocation of resources. Thinking in terms of a single commodity or service, it would be rational to employ a marginal (additional) unit of resources in its production so long as the benefits derived exceeded the associated costs. Elementary economic theory assumes that, after a point, as the production and consumption of a commodity increases, marginal costs rise and marginal benefit falls. Maximum benefit is thus defined by the point at which marginal benefit equals marginal cost (see box 5–2, figure 10). To go further would mean that marginal cost exceeded marginal benefit, and so the benefits to be derived from the final unit of production would be exceeded by the costs (see box 5–2, figure 11).

The optimum allocation of resources will also depend on how they are used. In practice resources are capable of alternative uses, and the total benefit derived from available resources will be affected by their allocation between these uses. Optimum allocation is represented here by the condition in which it is not possible to increase total benefit by switching resources from one use to another. Marginal benefits from resources would be the same in all uses.

A pollution tax or charge is in effect a negative subsidy. The market mechanism having failed to take all relevant costs and benefits into account, the tax is an example of government intervention to counter the distorting effects of the unplanned subsidy. Alternatively the process might be seen as one of simulating the market mechanism. Ralph D'Arge and James Wilen point out that economists commonly use one of two arguments to justify pollution charges.[2] The polluting firm may be regarded as a producer of joint products, a traditionally market-traded commodity and another, unpriced discommodity, namely pollution. Or, one could say that a common-property resource, the assimilative capacity of the environment, was being used without payment. Whether pollution represents a product or the use of a resource, a pollution tax is a phantom price intended to supplement normal market information.

In market theory, a market price is an automatically generated, *natural*, *true* indicator of relative scarcities and preferences. So, the price of oil would be taken to represent its true scarcity, and the real strength of consumer preference for oil over other products. In practice, as we saw in chapter 5, prices may to some extent be administered by giant corporations. Also, as the example of externalities shows, it cannot just be assumed that the market price is the *right* price. This raises some interesting questions about the role of taxes in environmental policy. At one extreme it might be held that market prices are the only data which should legitimately inform decision-making. Taxes would be seen as arbitrary, distorting interventions by governments. Alternatively, and most commonly among economists, pollution taxes would be regarded as supplementary information which simulates market mechanisms in situations where the normal market mechanism is inadequate. Both these views distinguish between information in the form of a market price and a tax, and ascribe primacy to market prices. A third position, however, might be that any indicator is going to be imperfect, either market prices or pollution taxes. So there is an implied case for a theory of information which does not stress the difference between them. It is true that the level of a pollution tax is set by a central authority and is to some extent arbitrary. However, the same can be said about market prices in the real world of giant corporations and externalities.

Of necessity, then, the level of any pollution tax is to an extent arbitrary. But in principle the tax represents those costs which have not been considered in the course of the normal market transactions associated with the production of a commodity (see box 5–2, figure 11). So how can the extent of the divergence between private and social costs be measured? In other words, what value should be placed on the disbenefits arising from spillovers? The level of pollution tax is arbitrary to the extent that these measurements cannot be made accurately. But some attempt must be made to place values on items which, for the very reason that they are not readily valued, fall outside the sphere of normal market transactions (see box 6–2). The choice is between implicit or explicit valuation, rather than between valuing these external effects or not valuing them. If no pollution charge is levied, although external costs are presumed to exist, then by implication a zero valuation has been placed on them.

Box 6–2   *Pricing the Environment*

Various approaches are possible. One is to observe other related features of the market. The disbenefits of aircraft noise, for example, might be reflected in the pattern of house prices around the airport. Also the political process might be seen as a market in which conflicting interests compete and are resolved. Alternatively, surveys could be undertaken to discover how much people would be prepared to pay for environmental programmes. Unfortunately none of these approaches is as straightforward as it might appear. House prices will reflect the presence of many other variables apart from aircraft noise; the relative influence of different groups in the political process may not be proportionate to the justice of their case. Can the *willingness to pay* data be accepted as an accurate reflection of preferences, or will it tend to under-represent those with low incomes? For such reasons as these, a decision based on cost–benefit calculations cannot be termed objective. Whatever calculations have gone into it, the decision will also reflect political pressures and governmental judgement. This element of judgement is inevitable. No matter how accurate the measure of the sum of individual preferences, it remains just that: a summation of *individual* preferences. The awkward problem of external, or *social*, costs and benefits is not resolved.

These problems of ascribing value to costs and benefits which fall outside the market mechanism are common to all types of environmental policy, taxes, subsidies, standards, etc. But although pollution charges share this problem, many advantages are claimed for this approach. Essentially these advantages amount to the fact that pollution taxes are thought to involve relatively little departure from the ways of the market. Also, once in operation they should be efficient in terms of the notion of optimality outlined in box 6–1.

A pollution tax internalises what was previously an external cost, and thus increases the accounted production costs of the firms concerned. The effects of such a cost increase will depend on the competitive conditions in the industry—that is, whether there are many competing firms or just a handful. Initially, books on environmental economics assume a competitive model. In a desire to reduce its profit margin, or increase its prices by as little as possible, each firm will consider the best response to the tax. Would it be cheaper just to pay the tax and carry on as before, or avoid some or all of the tax by adjusting operations? These adjustments could take various forms:

(a) effluent treatment;
(b) relocation of plant;
(c) changed production methods or materials;
(d) reduction in output.

In other words, the existing pattern of economic activity has arisen within a given framework of price signals. Alter the structure of prices, by introducing a pollution tax for example, and reactive changes can be expected (see box 1–7). Thus some firms might find it cheaper to invest in effluent treatment plant, rather than pay the tax on a particular pollutant. In that a large outlay may be required, this approach is more likely to attract the larger firms. If the potential gains are large enough, this could encourage mergers. Alternatively, specialist firms or joint ventures might be created to exploit any economies of large-scale effluent treatment. Some firms of course might prefer to carry on polluting and pay the tax. The theoretical outcome, however, would be that the simulated market mechanism would promote some effluent treatment. This would be done by the most efficient methods, and up to the point at which the private costs of treatment became equal to the social benefits of abatement as represented by the tax.

The tax might vary from place to place (depending on factors such as population density, prevailing wind, river conditions) or from nation to nation. In this case, another option open to a firm would be the relocation of its polluting activities. It might be objected that pollution is pollution, is pollution. . . . However, it is at least possible that the social costs of pollution do vary with natural conditions and population density, or even from nation to nation. A relatively underdeveloped nation or region might place a lower valuation on the social costs of pollution than an already industrialised area. In this case there would be theoretical reductions to be made in social costs by relocating the pollution.

Another way of avoiding the tax may be to use alternative production methods or materials. Often several industries share in the creation of a particular pollution problem. It is possible that some of these might be able to make such a switch quite readily. The amount of pollution could then be reduced at the lowest cost, in effect by *allocating* the unavoidable pollution to those industries least able to avoid it.

Finally in this list of adjustments to a pollution tax, there is the possibility of reducing the output of the pollution-generating end product. Individual firms might cut their output to arrive at a new equilibrium position now that a tax has been imposed and their costs raised. Also, if increased costs are reflected in higher prices, sales may fall independently of the manufacturers output plans. As a result some firms might also be persuaded to leave the field entirely, and potential new entrants might be discouraged.

To summarise, the theoretical attractiveness of the taxation approach to pollution abatement lies in its compatability with the belief in the efficiency of the market as a resource allocation mechanism. By some means (see box 6–2) an estimate is made of the external cost of pollution. This is represented by the tax and industrial decision-makers are free to adapt to the new structure of costs. A certain amount of pollution will be left after these adjustments, in that the marginal cost of pollution abatement will rise as higher levels of cleanliness are approached. After a certain point the cost of avoiding a further unit of pollution would exceed the amount of the tax—the marginal private costs of abatement would exceed the marginal social benefit. But the attraction of the tax approach in principle is that the abatement which did take place would be undertaken by those firms who could do it at least cost. The inefficient abaters (who found it more costly to reduce pollution by a given amount due to the nature of their business, location, etc.) would effectively be allocated the scarce licence to pollute. Put another way, the amount of resources used in bringing about the desired reduction in pollution would be minimised. Society would get the most abatement possible for its money—an abatement best buy.

So far, this section on pollution taxes has focused mainly on the advantages which are claimed for this approach to pollution abatement, and some of these will be further highlighted later when discussing alternative policies. However, there are problems and objections associated with pollution taxes. The last paragraph, for example, implied that some conservationists might object that it was simply putting a price on pollution, not stopping it. Hopefully, having read this book those holding this opinion would realise that this was not a valid objection. Pollution is rightly regarded as bad, but it does not follow logically that any and all abatement is good, because it depends on the relationship between the benefits and costs of abatement. But although it is in this sense invalid, the conservationist objection does illustrate the problem of setting the right level of pollution tax. Those with a strong belief in conservation, and those living near a chemical plant, may place a higher value on pollution abatement than the community at large. This also illustrates the possibility of political opposition to the introduction of a tax, rather than some other policy such as a ban on pollution. Industry might also oppose such a tax, or at least a high level of tax, on the grounds that it raises costs. Other interest groups could well share this objection, such as employees and local government authorities, particularly if the industry dominated employment in a region and was already in decline. In export industries there could be similar opposition to the introduction of a pollution tax. If other countries had no such tax, or a lower rate, export prospects might be endangered.

Such points are not valid objections to the principle of a pollution tax. Theoretically, what they represent are differences of opinion on the values to be placed on the costs and benefits of pollution abatement. But in practice, although amounting to the same thing, it might be more realistic to see them as the expression of sectional interests. It has been said that the reason why pollution taxes are not used is not because they would not work, but precisely because they *would* work. The benefits from the imposition of a pollution tax would be diffused over the whole community. Any adverse effects or inconvenience associated with the tax would, however, be concentrated on specific, organised and (for this reason alone, if no other) politically effective interests. In addition to these realities of political and economic structure, there is the more general obstacle to pollution taxes, namely that they imply that the environment is a communal resource (see the final paragraph of this section), an idea which would run counter to the dominant way of thought in a capitalist economy.

In addition to these objections there is also the problem of the cost of operating a pollution tax, although this may be lower than with some alternative policies. In order to set the correct tax level some research must be done to estimate the social costs of pollution. Legislation and administrative rules would have to be formulated to define the *unit of pollution* on which the tax would be levied (see box 6–8, p. 128). Finally, there would need to be some monitoring of industrial processes to assess each plant's tax liability (see box 6–9, p. 128). However, information would not be needed on a firm's previous level of pollution or on the most effective abatement technology, either of which might be required if a subsidy rather than taxation policy was followed (see section 6.1.2). It is because of these problems of administration and information costs, as well as the political opposition mentioned earlier, that alternative policies are often adopted. A simple pollution standard (see section 6.1.3), for example, is far from ideal in terms of the theory of optimal resource allocation, but it is more practical.

6.1.2 *Subsidies*

Technically at least, an alternative to penalising a company for creating pollution, is to offer it a reward for not doing so. Pollution represents an external cost which is borne by the community at large rather than the firm responsible. It would be quite logical then for society to calculate how much it would be worth to avoid these costs which are being imposed on it. Wherever the social cost of a unit of pollution exceeded the private costs which would be incurred by industry in avoiding it, it would be in the community's interest to cover the cost of abatement by a subsidy to industry. The benefits gained from the resulting abatement would be greater than the cost of the subsidy. Industry will have lost nothing and the community will have gained. The subsidy could be attached to abatement equipment, to particular geographical locations or to each unit of pollution avoided—by comparison with a firm's standard level of emission (see figure 12).

In fact there are theoretical, practical and political difficulties associated with such an approach. Take, for example, the subsidy attached to the reduc-

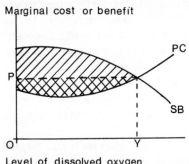

Marginal cost or benefit

Level of dissolved oxygen

Figure 12 Subsidising the Polluter (showing the marginal cost of—or benefit from—pollution abatement). This represents a simplified case where there is only one polluter and a fixed subsidy is paid for every unit of dissolved oxygen in the river up to a certain level.

SB = Marginal social benefit of pollution abatement. This rises as the water becomes good enough for fish life and then falls, suggesting that lower value is placed on increases in water quality beyond this point.

PC = Marginal private (polluter's) cost of pollution abatement. This falls for a while, reflecting economies arising from effluent treatment on an increased scale. After a point, however, additional water quality becomes increasingly expensive to achieve.

A fixed subsidy is paid of OP per unit of dissolved oxygen up to the level OY. The social benefit of pollution abatement up to this point exceeds the amount of subsidy by the area shaded. The polluter has an incentive to reduce pollution because the subsidy exceeds the cost of abatement by the area cross-hatched.

tion of pollution below a standard level. For each firm it would be necessary to know the standard level of emission and it would be costly to acquire this information as well as difficult to assess its accuracy. In the case of a preliminary period of monitoring, or a questionnaire, there would be the danger of exaggerated levels of pollution being engineered or claimed.

Unlike pollution charges, the subsidy payment for reduced pollution would not discourage entry into the industry, or raise costs and prices and thus lead to a lower level of output. The straightforward subsidy on effluent treatment plant has the additional disadvantage that it does not provide an incentive to change production methods or materials to give a lower pollution–output ratio. This point illustrates the general theoretical deficiency of the subsidy, compared with the tax approach. The information requirement for a pollution tax is simply an estimate of the external costs or damage arising from pollution. Industrialists are then free to choose between various possible responses to the tax, with the result that the best use is made of the resources devoted to pollution abatement. The subsidy, on the other hand, simply promotes a single response—the installation of treatment plant, or relocation—depending on how eligibility for the subsidy is defined. At best, in the case of a payment for each unit reduction in pollution below a standard level, a polluter will only be responsive up to the point at which the cost of avoiding a unit of pollution rises to the level of the subsidy. After that it would be cheaper to continue polluting and forego the subsidy. Of course, the same sort of equilibrium point exists in the case of a tax, when it becomes cheaper to pay the tax than incur the cost of avoiding it. In this case, however, the tax is still paid on the remaining pollution.

So technically a subsidy is inferior to a tax. It is administratively more difficult to define eligibility for, and pay, a subsidy than collect a tax. The information requirements are also greater for the subsidy in order to decide which behaviour it is best to subsidise—relocation, effluent treatment or the use of alternative fuels or materials. There are also political aspects to the choice of pollution control policy. As already seen, industrialists, employees and local authorities might oppose a pollution tax. With a subsidy, others might make the general objection that the polluter should pay, rather than be paid. The explanation of this view is probably to be found in attitudes towards the environment. The subsidy approach implies that the right to use the environment lies with industry and that the community must pay for any alternative uses. Thus, the chemical plant would have the right to use the river for waste disposal. If local citizens wanted to use the river for recreational purposes, they would have to buy the option by paying the chemical company to reduce the flow of effluent. With a pollution tax, on the other hand, the implication is that the environment represents a community resource and the chemical firm should pay to use it for waste disposal.

6.1.3 *The Role of Law and Government*

Chapter 5 made the simple point that environmental problems are 'essentially man-made—they are things we do to each other or to future generations.' The relevance of the law is thus quite clear because, 'the greater part of any national system of law is concerned with the relationship between individuals and/or associations of individuals and the organs of government.'[3] So, as Professor Denman warns, an examination of law relating to the environment reveals 'an extensive web, ready woven, a gossamer of statutory tests and case law stretching back into the past when in the law of torts neighbour sought redress against neighbour for nuisance of various kinds, when legislators focused their attention on housing, on factory conditions and on the neglect of husbandry'.[4] Today, everyone will be aware, at least vaguely, of this maze of legislation whose growth has been associated largely with the process of industrialisation and urbanisation. Every aspect of the environment, broadly defined, is in some way a subject of the law—land use (the zoning of activities, conservation areas, green belts, national parks); all aspects of production such as agriculture, mining and quarrying (including tipping and dereliction), industry (industrial location, conditions of work), transport (the siting and use of roads and airports), distributive trades (conditions in shops, consumer protection); the built environment (housing, building regulations, historic buildings, the siting of offices).

Perhaps the archetypal environmental problem is pollution and this can be used to illustrate some general points about law and the environment. D. A. Bigham sets the scene when he reports: 'At present the law relating to the various forms of pollution is complicated and confused. The remedies available vary and overlap between those offered by the courts at common law and in equity and those offered (frequently concurrently) by statute or by Ministerial Rule and Order, or even by Ministerial 'advisory circulars'.[5]

The most obvious application of the common law relates to nuisance, that is where a person is unreasonably and unnecessarily inconvenienced by some activity of his neighbour. But, as Allan D. McKnight observes: 'On the whole, the law of nuisance is full of so many intricacies as to be almost inapplicable to the circumstances of a modern industrial society in which a single householder (or a small group) is attempting to defend himself by legal action against the invasions of the use of his land or his enjoyment of his home by neighbouring industrial enterprises.'[6]

To elaborate, although the individual may seek a common-law remedy, in practice there are many obstacles to the effective use of common law as an anti-pollution device. The individual would need to possess the knowledge and the initiative to start proceedings and not be deterred by the uncertain cost and uncertain outcome. The outcome would be uncertain because the individual would have to prove, with noise for example, where the offending noise was coming from (his only hope would be in the extreme case of a single, stationary source), and that he was more than averagely affected. (He could not simply claim to be more sensitive than other individuals, and neither could he object successfully to the typical noise of his neighbourhood—an industrial area is naturally going to be noisier than the countryside, and it is the local area, not some objective standard, which is taken as the norm.) Another source of uncertainty for the individual would be judicial attitudes. The Noise Abatement Act 1960, for example, makes 'excessive' noise a statutory nuisance. The difficulty of defining 'excessive' favours the noisemaker, particularly as the social usefulness of the industry creating the noise is taken into consideration in making a judgement.[7]

In practice, therefore, action has to depend on local and central authorities. David Harris sums up the position on law and pollution generally when he says that in the United Kingdom 'reliance is placed mostly upon legislation establishing administrative machinery for the control of air pollution, leaving common-law nuisance very much as a 'long stop'.[8] But in the field of statute law too there are deficiencies, as writers on law and the environment point out:[9]

(a) There are gaps or areas where the law requires strengthening, for example, on sulphur dioxide, pesticides, vehicle noise.

(b) Where standards are set, as with noise in decibels, the limit tends to relate to existing rather than desired levels.

(c) Taking noise as an example again, the damage is already being done by the time the law is invoked (except in the case of zoning laws and building codes).

(d) The expensive machinery associated with a noisy or polluting activity has already been installed, therefore when commercial considerations enter into the judgement there is a bias in favour of the status quo.

(e) The zealous application of existing powers, for example, with relation to water quality, would impose (unacceptably) high costs on local authorities and industry.

(f) There are problems of inadequate monitoring and enforcement, as in the case of air pollution, the Alkali Inspectorate being understaffed.

(g) Penalties are relatively small.

Apart from extending the coverage of the law, other improvements could be made. Information, on pollution levels and sources, could be made available to the public, which would help pollution victims make a case. Also, the law might be changed so that a community suit could be filed, rather than each individual having to fight a separate legal battle.

The use of law in the way just described represents a direct attempt to resolve environmental conflicts outside the market framework. However, it is also possible to think of legal and institutional innovations which achieve their effects indirectly—by removing obstacles to normal market transactions. As suggested in chapter 5 (p. 96) the fundamental problem is the common-property character of rivers, seas and the air. An example given there was the use of a river for waste disposal as if it were costless. It appears costless because those who would be affected by the pollution do not have to be compensated. They do not have the normal property rights which form the basis of a market transaction and their interests are thus not represented effectively by market mechanisms.

In principle, there are several possible responses to this situation. Essentially, the problem is that some costs are not being considered, with the result that resource allocation is being distorted. Taxes and subsidies illustrate the general solution—to make sure that the hidden costs do appear in decision-makers' accounts. In the case of a common-property resource, this might also be done by assigning property rights, either to the instigator or the victim of pollution, and then facilitating transactions between the parties.

Consider the implications of the chemical company owning the entire river. It would still be available for the disposal of their waste, but there would now be other factors to take into account. The river would represent a potential earning asset: from the sale of water to industry, power stations and water authorities; licensed waste disposal by other firms; licensed fishing and boating. This would have the effect of internalising the previously external costs of pollution. There would now be a cost to the chemical company in the sense of the earnings foregone if their pollution of the river reduced income from these other uses.

The existence of property rights is a necessary but not sufficient condition for the operation of a market. Another relevant factor is transaction costs. The different interests must be able to negotiate with each other, and the obstacles to negotiation must not be so great that the transaction costs exceed the potential benefits. In many cases, particularly in the field of nuisance, this is often the problem. Each individual may have the right to reasonable peace and quiet, but it is not feasible for the creators and victims of noise to negotiate. This is a particular problem in consumer–consumer pollution where there are numerous contributors, for example to noise or to car exhaust, and also many victims. It also occurs, however, where there are many victims but only a few polluters. There are difficulties in identifying the individual polluter's contribution, as well as obstacles to effective action on the part of individual victims.

Because of these difficulties about applying market or market-like mechanisms to pollution, a more simple, but theoretically less ideal, approach is often used. The interests of the community in clean air, water, etc. are expressed in regulations on pollution standards. This should still involve calculations of costs and benefits, otherwise a standard could be set where the benefits from reduced pollution were outweighed by the costs of achieving it. This problem of the optimal level of the standard is paralleled when it comes to applying the standard to particular sources of pollution. Ideally, in order to minimise the costs of abatement the total reduction in pollution which it is desired to achieve should be distributed between the contributors in such a way that the marginal cost of abatement is equal for each of them. In this case, the total costs of abatement could not be reduced by arranging for one polluter to cut his effluent level by a little more, and another by a little less. Those who can avoid pollution cheaply will reduce it by more units than those for whom the costs of abatement are higher. From the viewpoint of efficiency, it would be wrong therefore to insist that each polluter reduce pollution by the same amount. But this would mean an individual standard for each location, involving high administration and information costs—the usual problem. So, with inadequate information, the standard will at best be a negotiated one (between the standards authority and the firms concerned), or at worst be completely arbitrary. In principle, however, a reasonably accurate standard is better than regulations about the installation of treatment plant. It leaves room for a variety of responses on the part of the polluter in order to minimise the costs of compliance with the standard. (A ban is an extreme example of a standard—a zero level—where by implication the damage is so great that it must be avoided whatever the cost.)

Even apart from these reservations about its efficiency in economic terms, the administrative approach to pollution control via regulation and standards is still not as straightforward as it might seem (see box 6–3).

In theory, one way of allocating efficiently the limited emission of pollutants which a standard represents is to market emission rights. In effect, ownership would be created in a common property, such as the air. This ownership would be assigned to, and marketed by, the state. At an auction of licences those polluters for whom there was no cheaper method would buy the licences. The community, assuming adequate 'policing', would achieve its

Box 6–3   Environmental Regulation in the United States

'There seems to be an assumption that regulation acts simply and directly, and that the issuance of a rule or an order by an administrative agency results in the achievement of the mandate and the purpose of that rule or order without any complicating consequences. This assumption is not to be found explicitly in any discussion but seems to be implied in most of the literature. To say the least, this assumption is uncritical, naïve and unrealistic'—D. T. Savage *et al.*, *The Economics of Environmental Improvements* (1974).[10]

Regulatory agencies are common in the United States where their defects have been well documented by various Presidential Commissions, such as the Ash Council, 1971. This report pointed critically at three characteristics of regulatory bodies:

(1) Regulatory commissions, consisting of six or seven coequal commissioners, lack clear lines of responsibility; this leads to shared indecision, unaccountability and administrative inefficiency.

(2) Instead of being imaginative policy-makers in an area of complex and changing problems, the commissions tend to become passive semi-judicial bodies.

(3) The overlap of responsibility among different agencies results in fragmentation and inflexibility. In the field of transport, for example, there is the Interstate Commerce Commission, the Civil Aeronautics Board, and the Federal Maritime Commision.

Similar criticisms are already being levelled at the Environmental Protection Agency. (Formed in 1970, it encompasses several earlier agencies, and combines within its mandate the main provisions of existing water quality and clean-air legislation). It is said to be acquiring a judicial atmosphere; the enforcement processes are slow, cumbersome and costly; firms and municipalities have been able to delay or avoid responding to E.P.A. orders.

The Ash report suggested that the effectiveness of regulatory commissions could be improved by making organisational changes—for example, having a single commissioner, and creating a separate administrative court to review the commission's decisions. Others point out that this would not alter the (political) environment in which regulatory decisions are made. For example:

(a) The level set for a pollution standard is the outcome of a political process. The individual states compete as hosts of industry, and may therefore tend to resist stringent environmental standards. Federal administrators, being dependent on the political support of the states, must therefore bargain and compromise. It is thus possible to find federally approved standards being set in a particular state at a level which permits an increase in pollution. For example, the 1968 Illinois standard on dissolved solids was ten milligrams per litre higher than the level in the already polluted Lake Michigan, and the state's cyanide standard was two and a half times existing levels.

(b) Regulatory agencies are remote from the public, and what public interest there is may be blunted by the technical nature of pollution control. Perhaps as a result of this, the regulators tend to identify with the regulated, with whom they have more contact and who are better organised to express their interests.'

(This exposition is based on chapters 9 and 10 of *The Economics of Environmental Improvement*—see introductory quotation—where there can be found a fuller discussion of environmental regulation and the history of U.S. federal environmental legislation.)

---

desired standard of emission at the lowest cost. Other potential polluters, who could more readily do so, will have installed treatment plant, or will have changed materials, methods or final products.

The general aim of the policy options so far mentioned has been to *internalise the externalities*. Taxes and subsidies bring hidden costs and benefits into account; the extension of property rights widens the area of interest of decision-makers; the provision of information and a legal-institutional framework facilitates negotiations where property rights already exist; the penalties arising from effectively policed standards and bans represent a charge on pollution. It should be apparent, however, that the demands of the theory of optimal resource allocation, that *all* costs and benefits be taken into account, are not so easy to meet in practice. As we have seen, obstacles to transactions between creators and victims of pollution are often insurmountable. The resort to government action, in the form of a tax, subsidy, standard or the extension of public ownership, *appears* to solve the problem by making it internal from the viewpoint of the community or nation. There is still the difficulty, however, of finding the right value to place on the benefits and costs involved. That a problem is made technically internal, does not alter the fact, most obviously in the case of a whole nation, that the different interests of many groups of people have to be resolved.

Finally, ecological problems are no great respecters of national boundaries. Pollution of the atmosphere and seas can have transnational effects, as vividly illustrated by the legendary penguins of Antarctica with the high concentration of D.D.T. in their tissues. Thus in some instances to internalise the problem would involve a global perspective. Having surveyed the difficulties of simulating, facilitating or replacing the market mechanism within a nation, the even greater problems presented internationally can be imagined. Of course, with numerous nations involved, the administrative difficulties of any policy are multiplied. More fundamentally, values and attitudes concerning environmental issues are likely to vary widely with national differences in economic and social priorities, levels of development and environmental conditions. For example, an already industrialised, densely populated country living at peace with its neighbours is likely to be much more sensitive to pollution than a sparsely populated, developing country which feels threatened by neighbouring states. It would thus be extremely difficult to gain universal acceptance of a particular pollution tax rate. Countries in urgent need of industrial development and imported capital, or with a balance-of-trade problem, would wish to attract foreign enterprises and keep domestic costs down by a low or zero level of pollution tax.

The effectiveness of an approach via international law and regulation is equally problematical. Apart from the attendant problem of enforcement, ready agreement cannot be expected. In the case of the law of the sea, take, for example, the issue of defining boundaries relating to territorial rights, fishing rights, the sovereign right to explore and exploit mineral resources, and the boundary on the sea bed for purposes of disarmament and arms control.

Geographical, technological and political factors will shape a nation's response to these matters. The extent of territorial water may have particular strategic and political significance for some countries. Similarly, the exclusive rights to fish in certain areas may be of profound economic and political significance—as in the case of Iceland. Differences in proximity to offshore resources and access to the necessary technology can be expected to influence a country's attitudes on the extent of the 'economic zone': that is, the area in which a littoral state has exploitation rights. In the same way, the positions adopted on the demilitarisation of the sea bed will clearly be influenced by technological, strategic and political factors.

Even in the sphere of marine pollution, where less disunity might be expected, the advance of international law has been slow. There is no system even of registration, let alone regulation, of dumping in the ocean. International law so far deals only with marine pollution caused by radioactive materials and oil from ships; and the drafting of the relevant provisions is in such general terms as to make the laws not effectively binding.

6.2 Resources, Population, Technology and the Social Environment

Pollution is only one aspect of the environmental problem, although it is probably the one of most immediate and widely appreciated concern. It has figured so largely in this chapter because it provides the best vehicle for illustrating the application of different policy approaches. The remainder of the chapter will look briefly at the possible implications of these approaches in some other major areas—resources, population, technology and the social environment.

Once again, the resources problem can be expressed in terms of the divergence between private and social, or between present and future, interests. The difficulties of estimating the social interest have already been referred to, and the problem of defining the interests of future generations is even greater. Assume, however, that resources are being used too quickly by either of these tests. A tax on, say, copper, should have the effect of stimulating users to employ it more efficiently, to recycle on a larger scale and to find more abundant alternative materials (which would bear a lower tax). A tax on copper products, rather than on the material itself, would have similar effects—cutting sales, output and perhaps profit margins, in which case greater efficiency might be promoted.

Subsidies could also be used to guide decisions on resource exploitation. To overcome immediate and short-term scarcities, more abundant materials, or exploration activities, might be subsidised. In the longer term of course, such policies would not act to conserve resources, but this is inevitably the case with any substitution of one non-renewable resource for another. The use of subsidies by a nation to promote more resource-efficient manufacturing technology would help to conserve resources—unless the savings were used by other countries or to expand production levels. The subsidisation of recycling and of recycling technology would be subject to the same possible limitations. Also, remembering that nothing is for nothing, recycling amounts to the substitution of one resource for another. The process always involves energy, for example, and recycling is not a magic wand with 100 per cent efficiency. The returns will vary, so if energy is extremely scarce and the conversion process technically inefficient, they could be small or even negative. Recalling what was said earlier about the role of prices in decision-making, such calculations can only be accurate if the prevailing prices which enter into them are accurate reflections of relative scarcities. If for some reason (such as a government subsidy on fuel designed to keep industrial costs and export prices down) energy prices were artificially low, then recycling decisions would be distorted.

As in the case of pollution, the problem is how to bring into account the costs and benefits which are external to the decision-makers involved, such as the interests of society as a whole or of future generations. Altering the structure of costs and prices is one approach. Another is to widen the area of concern of decision-makers by extending property rights. For example, where several users were competing to use a dwindling resource, the effect might be to accelerate its depletion. As the exhaustion of reserves approached, exploitation would become more frantic as each user tried to secure as much as possible. If mergers and takeovers reduced the number of competitors the possibility of agreement on a planned rate of exploitation would be increased. To go even further, if one company owned all reserves then a company policy on the management of the resource would in effect be a global policy.

Even then, the resource would be managed in the interests of the company which could diverge from the wider social interest. A logical next step to internalise the social costs and benefits would be communal ownership. The resource would be owned by the state and managed in the interests of the community at large. But, as we saw earlier, there is still the problem of assessing this 'social interest'. Also, how will the state evaluate the interests of future generations? This is no easy matter, particularly when it is known that governments, as much if not more than most of us, are primarily concerned with more immediate returns.

A half-way house to state ownership is regulation. Conceivably, the interests of society as a whole, and of future generations, might be reflected in regulations which limit the power of those who own or control resources. Thus there could be legal restrictions on the rate of resource utilisation. These might take the form of limits on the amounts of minerals mined, fish caught, etc. or the quantities imported, or, in the case of a producing nation, exported.

Of course, the promised neatness of regulations establishing the rate of resource depletion is purely theoretical. In practice resources are often concentrated in a few countries or may be controlled by multinational corporations. In neither case will it be possible for any individual consumer nation to control resource use. Furthermore, some resources such as fish or whale stocks are common-property resources falling outside any nation's jurisdiction. It might be added here that in one sense, in circumstances where reserves are running low, any resource has something of a common-property character: if one country does not use it another will.

In this situation, apart from the balance of payments considerations which apply to an importing country, what incentive has a consumer nation to cut back on resource use? Levels of income and employment and, in the extreme, social and political stability, may be seen as dependent on the continued growth of economic output. A nation's international military and political fortunes are also likely to be affected. Logically some multilateral de-escalation of resource use, rather like the disarmament process, would be needed. The difficulties are obvious. Even if all countries agreed to participate, there would be marked differences of interest, depending on levels of development, economic structure and social and political conditions. Also, what would constitute a fair formula for the cuts? Would each consumer nation take a cut proportionate to existing usage, or should the major consumers make higher percentage reductions? In either case would this not bias the distribution of resources essentially to those countries already using them? What about nations hoping to develop manufacturing industry? Perhaps they would prefer not to see cuts of whatever proportions from given consumption levels. Instead their proposal might be to start with the total permitted consumption and allocate it equally to all countries. An individual nation could then use, stock or sell its share. The obvious implication of all this is that the control of essential scarce resources be vested in an international body. Needless to say, this does not short-circuit the truly enormous practical political obstacles to the global management of resources. The difficulties just described would have to be overcome before any such body could operate effectively (see box 6–4).

In principle at least, the same sort of approaches could be taken towards population. The practical policy options available to a government will depend on the prevailing social, economic and political conditions, but essentially they could act on the willingness, together with the ability, of people to limit their families. Willingness might be influenced by increasing the 'costs' associated with children, reducing or replacing the 'benefits', or by changing attitudes to family size within a given structure of costs and benefits.

For example, in Britain, the view has been expressed that tax allowances, cash, and indirect benefits like state medical and educational services, constitute a subsidy to families. Some writers have therefore proposed that these be reduced or abolished. The expectation is that family size would be reduced as a result because the effects would be like introducing a tax on children. This implies that financial considerations are important in determining willingness to support a family. In fact motivation may well be more complicated. Recognising this, a more sophisticated approach would be to identify the benefits derived from a family and provide substitutes for some of them. Thus in a developing country, where a large family may provide

Box 6–4   International Resource Management

Although very far removed from the concept of global resource management in the fullest sense, the 1974 Washington energy conference illustrates the problem in its more limited contemporary setting. Some consumer countries argued for a multilateral response to the oil crisis—because two-way deals would bid up the price; could be threatened by any future political squabble; could result in competition to supply arms for oil; and would undermine western political unity. Others claimed that the formation of an oil consumers club would be provocative, that in any case conditions for a unified approach within Europe were not present, and also that U.S. interests in the Middle East diverged from European interests.

In the case of other raw materials, the interests of consumer nations would be even more diverse. '. . . this seems to point away from a multilateral, cooperative framework of policy. And yet, the example of the particular cooperative arrangements for oil set in train by the Washington conference is likely to be in existence for some time. Provided that these arrangements do not conspicuously fail, they may well encourage a general disposition to arrange other commodity problems under Washington-type procedures . . . institutions which currently exist—the European Community, the multinational companies, the United Nations and the rest—will be, for better or worse, and with no more than minor exceptions, the only institutions which can be relied on to contribute during this critical period (before Third World attitudes harden and the impetus of consumer co-operation on energy is lost) to solving the political and economic problems of adaptation to a new resource era'—P. Connelly and R. Perlman.[11]

economic security, changes in land tenure and the organisation of agriculture could remove the need for this type of security. Family plots might be replaced by communal farms supported by government economic and social services. This sort of approach, based on an analysis of motivation, is likely to be the most effective. Whatever the approach, demographic change will only occur gradually, but if important elements of motivation are ignored then the withdrawal of cash subsidies or free services, the supply of contraception information, devices and incentives, or propaganda campaigns are less likely to be effective.

No one cause can be singled out to account for pollution, resource depletion and the character of our daily lives. The growth of population and production are relevant factors, but so is scientific and technological development. The pace of such development has been accelerating, and present-day environmental problems may themselves spur further developments—such as nuclear fission processes to bypass the shortage of conventional fuels.

In recent years, by dispensing contracts and funds, governments have influenced the direction of some research and development. Yet the powerful engine of scientific and technological development can still be described as fragmented. Government departments, scientific and industrial enterprises, each with a particular interest, have contributed to the complex and undirected growth of technology which shapes our present and constrains our future. Legislature can undertake what amounts to retrospective assessment by controlling or even banning a technology when its harmful effects become known. There is a growing interest, however, in the integration of the ongoing process of scientific and technological development with broader social objectives.

This would be a far more complicated matter than simply establishing an Office of Technology Assessment (see box 6–5).

The evaluation and control of the whole range of actual and potential technologies presents both technical and political difficulties. Information and research requirements would be massive. Also, each technology which has been introduced, or is planned, will be beneficial to some group and may be supported by interested manufacturers, workers, consumers and perhaps the responsible government departments. To be effective, the assessment body must thus be able to claim that it represents the wider social interest, and the government must also have the power and the will to implement decisions based on the assessment. This process has radical implications. For example, would it be possible to involve the community in the assessment process or would it effectively concentrate the power to direct the development of science and technology in the hands of an isolated elite? In either case the increase in bureaucracy and socialisation in science clashes with the traditional scientific ethos of disinterested enquiry. But can science be left to the scientists? Finally, if effective, the process of assessment would lead to the rejection of some scientific and technological developments which would otherwise have been accepted. How would we react to the objections that the rejected project would have been commercially successful; that it would have

---

Box 6–5   *An Office of Technology Assessment*

The Office of Technology Assessment in the United States ... is designed to give independent analyses of technological matters being considered by the Congress. It is supposed to work chiefly by contracting out studies to universities and non-profit organisations. Studies can be initiated by congressional committee chairmen or at the request of the ranking minority member or a majority of members of any congressional committee. It is managed by a director and a board of management which have the authority to initiate technology assessment studies; they are responsible only to the Congress.

In carrying out its function, the Office of Technology Assessment shall:

identify existing or probable impacts of technology;

where possible establish cause and effect relationships;

determine alternative technological methods of achieving requisite goals;

make estimates and comparisons of the impacts of alternative methods and programmes;

present findings of completed analyses to the appropriate legislative authority;

identify areas where additional research or data collection is required to provide adequate support for the assessments and estimates;

make information, surveys, studies, reports and findings freely available to the public except where to do so would violate security statutes.[12]

---

created employment; that it would have enhanced our country's influence and prestige; and that vetoing it gives competitive or even military advantage to other countries? François Hetman suggests that 'Examples of such technological developments as the atomic bomb, weather modification, various military techniques, chemical agents and biochemical agents, etc. show that the existence of substantial hazards has not deterred such developments because optimistic realisations have been accepted at the level of sponsoring bodies and at the highest levels of decision-making process.'[13]

Although the debate on the environment is conventionally conducted in terms of pollution and resource depletion, the quality of life for the average citizen is determined by a host of less dramatic factors. If defined broadly in this way, the environment is influenced by many government policies to some of which the term 'environmental' may not normally be applied. Policies on housing, land use, transport and waste disposal and clearly relevant. They directly affect the physical environment. Rather more indirectly, but no less surely, government policies on the distribution of incomes, the provision of cash benefits or medical, educational and social services, also have a significant impact on the quality of life for the individual. The same can be said of areas of policy which influence the working environment such as legislation on issues like redundancy payments, unfair dismissal and worker participation in the running of undustry.

Recognition has grown that conventional performance indicators like profitability or gross national product, do not take proper account of pollution, resource depletion, the distribution of incomes and many important qualitative aspects of life. Consequently, proposals have been made that a new form of balance sheet be created which would be useful in clarifying policy choices. There is, for example, the suggestion that industrial enterprises should be subject to a *social audit* (see box 6–6) and that research be undertaken into the possibility of a system of national social accounts. Daniel Bell quotes from an article in the 1968 yearbook of the U.S Department of the Interior: 'Gross National Product is our Holy Grail . . . but we have no environmental index, no census statistics to measure whether the country is more liveable from year to year.'[14] Apart from indexes of air and water quality, such an environmental balance sheet would also need to include the social costs of technological change, measures of the extent of social ills such as crime, measures of the achievement of targets in housing and education, and indicators of economic opportunity and social mobility.[15] This may sound logical, but there are major practical problems of gathering the necessary information, and also great theoretical difficulties in measuring such elusive factors and combining them in an index.

Summary

Starting from the viewpoint that environmental problems are the result of resource misallocation, this chapter has explored the possibilities of influencing decision-making processes. For example, taxes, subsidies or regulatory agencies might be introduced to represent relevant information which has been left out of account. The practical use of this sort of approach to the con-

---

Box 6—6   *A Social Audit of Industry*

'... There must be, and be seen to be, an ethical dimension in corporate activity. Companies must in our view recognise that they have functions, duties and moral obligations that go beyond the immediate pursuit of profit and the requirements of the law.'[16]

From a checklist offered by John Hamble,[17] some possible headings in a social audit are—'the external environment'—relations with the community, consumers and shareholders; pollution; packaging;—'the internal environment'—working conditions; treatment of minority groups; industrial relations; education and training.

The pressure group Social Audit proposed to put a resolution at the annual meeting of Tube Investments Ltd that there be published annually a report giving comprehensive account of the company's activities in:

(a) the prevention of accidents and ill-health at work;

(b) the conservation of energy and natural resources and in pollution control;

(c) recruitment and conditions of employment for minority groups and women;

(d) military contracting;

(e) handling complaints about the safety, quality and durability of consumer goods and the guarantees and servicing arrangements provided for them.

In their report, Social Audit drew attention to the problem of obtaining the necessary facts: 'Until the disclosure of such basic information is required by law, comprehensive social auditing—measuring the total cost and contribution to the community of a business operation—will not be feasible.'[18]

---

trol of water pollution in the Ruhr area of West Germany is illustrated in box 6–7.

Section 6.1.1 discusses the theoretical efficiency of the taxation approach to pollution control compared with subsidies and regulation. Taxation represents a simulation of the market and allows decision-makers maximum freedom of response. The problem remains, however, of determining the correct level of charges in differing circumstances. The price of increasing accuracy is rising costs of information and administration. Box 6–8, which

---

Box 6–7   *Ruhr Regional Water Authorities*

There are seven large water resources Cooperative Associations called Genossenschaften in the highly industrialised and heavily populated area generally known as the Ruhr. They have the authority to plan and construct facilities for water resources and to assess members with the cost of constructing and operating such facilities.

They have realised major gains from viewing the waste disposal–water supply problem as one of a system character rather than solely as a matter of treating wastes at individual outfalls. They have made extensive use of scale economies in treatment by linking several towns and cities to a single treatment plant—in the case of the river Emscher they have, in fact, linked an entire watershed to a single treatment plant. They have in other cases induced waste recovery or changes in manufacturing processes by levying charges for effluent discharge based on quantity and quality of waste water.

Numerous adjustments, especially in manufacturing process design, are being made by the industrial plants in the Ruhr area as a consequence of the Genossenschaften's methods which force industry to bear at least a significant portion of the social costs of waste disposal.[19]

---

describes the procedure for fixing effluent charges for the Emscher river, gives some insight into these problems.

It can be seen that in practice compromises have to be made. Ideally, the system of pollution control should minimise the resource costs of achieving abatement objectives. Thus, although different instances of pollution may have the same physical effect, for example, on fish life, the costs of abating them (by treatment or changed production methods and materials) may be very different. The same might be true of identical pollutants released under different river conditions. To take all these circumstances into account, however, would greatly add to the costs of information and administration. So, in practice, less than ideal compromises have to be accepted, as box 6–9 illustrates in the case of the Ruhr.

In Section 6.2 we have examined the application of such measures as taxes, subsidies, regulations, etc. to other aspects of the environment issue such as resources, population and technology. In choosing the correct level of

---

Box 6–8   Ruhr Regional Water Authorities: Calculating Effluent Charges

'The question confronting Genossenschaften is how the diverse wastes produced by industrial enterprises can be assessed with an appropriate portion of costs. Very briefly put, the Emschergenossenschaft procedure is roughly as follows: (1) There is estimated first an amount of water necessary to dilute a given amount of waste materials subject to sedimentation in order that they may not be destructive to fish life under the conditions of the area. An amount of dilution water required by such materials in a given effluent is then calculated on that basis. (2) An analogous calculation made for materials subject to biochemical degradation but which are not subject to sedimentation. (3) The amount of dilution required under specific conditions in order that the toxic material in the effluent will not kill fish is computed by direct experimentation. (4) Certain side calculations having to do with water depletion, heat in effluent, etc. are made. The derived dilution requirements are added together for the effluent and form a basis for comparison with all other effluents. In principle, costs are distributed in accordance with the proportion of aggregate dilution requirements accounted for by the specific effluent'—Kneese.[20]

Box 6–9 Ruhr Regional Water Authorities: Information and Administration Costs

'In assessing the merits of applying 'peak load' principles to effluent discharges, the costs of determining variation in the quality and quantity of effluent over the relevant period must of course be considered. Presently, the Genossenschaften generally establish by sampling an effluent quality which is taken to be typical for the year. The quantity of effluent discharged during the year is ordinarily based on measurements reported by the plant. The costs of operating analysis sampling programmes along present lines would mount sharply if an effort were made to determine quality and quantity variation with sufficient continuity to permit peak load pricing.'

The Ruhr procedures can be criticised on technical as well as economic grounds. They are indeed recognised as less than ideal by the Genossenschaften but are generally defended as being readily understandable and relatively inexpensive to administer.[21]

tax or the appropriate standard to be applied, the same technical and economic difficulties are to be found. In addition, as the cases of resource rationing and the control of technology most clearly illustrate, even more formidable obstacles arise from the political, social and international implications of effective policies in these areas. All this raises the question: Will environmental problems respond to reforms in existing decision-making procedures, or are the problems to some extent built-in to present-day social, economic and political systems? Chapter 7 turns to the issue of more fundamental change.

CHAPTER 7

An Evolving Post-industrial Society?

Earlier chapters have summarised the different interpretations which have been given of the environmental problem as a whole, and explained some of the main constituent elements such as population, resource depletion and pollution. These environmental issues can only be seen in perspective if they are placed in the context of society. From the viewpoint of economists, a unifying feature of environmental issues is their connection with processes of resource allocation. It was argued in chapter 5 that energy and resources must have been applied to create any environmental effect. Effectively, a choice must have been made on the basis of some implied criteria or set of values. Given that a range of values is likely to exist in a society, then, on any given issue, such as the building and location of an airport, some interests must have prevailed over others.

These decision-making processes are undertaken in a society's political and economic system. If this system is thought to be characterised by fairness and rationality, then any environmental problems are likely to be seen as externalities—the outcomes of specific defects in the decision-making processes. In this case, the main thrust of environmental policy, as described in the last chapter, will take the shape of measures designed to counteract particular problems. The examples given were of taxes or subsidies aimed at reflecting broader social costs and benefits in market prices; the increasing role of 'communal' decision-making aided by cost–benefit studies and planning, and the increase in environmental legislation.

This approach seems logical, but is it adequate? Could it be that some combination of the scale of environmental problems, the pace at which they appear and the way they combine with each other, demand changes which go beyond the framework of contemporary society, requiring a reshaping of its structure? Is it possible that the problems are not external in the sense of being merely symptoms of some specific malfunction which, although serious, will respond to localised treatment? What if the malady is built in and requires surgery, organ transplants or even genetic re-engineering? If this is the correct analogy to apply to present-day society, it is possible that our conventional treatments may make matters worse. Also, even if the correct diagnosis was to be made, are we capable of taking the needed steps? In any case, will there be time for such a deliberate approach, or shall we have to adapt (or fail to adapt) to circumstances as the situation develops?

Although perhaps not peculiar to our own age, there is evidence for the existence of this sort of uncertainty. Arising from an awareness of the accumulation of problems, the accelerating pace of change and maybe the approach of the millennial year 2000, there is a widespread interest in the shape of future society and the path towards it. Two complementary and overlapping elements feature in the approaches adopted by writers on the subject. The first is an historical, scientific, objective method which tries to identify the forces which have shaped cultural change in the past, It asks what determines the nature of a society, is it the character of its technology, its system of social relations or its ideas—values, beliefs, modes of thought? Clearly these are interrelated, but various writers emphasise one or another. Contemporary futurologists illustrate the objective element when they attempt to identify and classify the chief characteristics of a society and project them into the future. The second element is more overtly subjective and when it is dominant the result is an 'involved' or 'humane' approach. The emphasis is on an author's perception of current or foreseen problems and his prescriptions for the changes needed if future society is to be consistent with his preferences.

This second approach can be illustrated by references to writers such as Ivan Illich, René Dubos, Rachel Carson and others. These also provide examples of points made earlier, such as the sense of uncertainty about man's present situation, an orientation towards the future and the feeling that major changes will have to be made in the fabric of society. For example, life in the West today, says Weisskopf, is permeated by unrest, uneasiness and a mood of pessimistic desperation.[1] Similarly, although he does not share it himself, Sir Peter Medawar, the distinguished British scientist, reports the presence of a 'sense of decay and deterioration, . . . of doubt about the adequacy of man, amounting in all to what a future historian might . . . describe as a failure of nerve'.[2] This uncertainty is probably related to an awareness of accelerating and uncontrolled change. Has revolutionary technology outstripped political and social imagination, so that an unparalleled transformation is being allowed to happen casually? This is the central point of Alvin Toffler's *Future Shock*—(1970)—that we are being accelerated into the future so fast that we cannot adapt to the changes. It is as if we are lunging forward in a train which gathers speed daily, but has no driver and no brakes.[3]

Where is this leading? Rachel Carson believes that 'The road we have been travelling is deceptively easy, a smooth superhighway on which we progress at great speed, but at its end lies disaster'.[4] This reinforces the point about the pace of change, but it also might suggest that we are being 'drawn' down the road to future disaster. Various hazards might appear, but they are negotiated with little difficulty. Speaking of the problems of American society, Ivan Illich says: 'Each is dealt with as a separate phenomenon, each is explained by a different report, each calls for a new tax and a new programme. ... Trying to bring about an era which is both hyper-industrial and ecologically feasible, they accelerate the breakdown ...'[5] Piecemeal responses made now close our future options. As René Dubos puts it: 'The state of adaptedness to the world today may be incompatible with the world of tomorrow.'[6]

Inevitably, writers with this view see the need for more fundamental change. The structure of society will have to change, but that will not happen until a disaster happens, and sooner or later one will, says Dubos.[7] Illich also believes that it would be folly not to expect in the very near future an event whose effects will overload our capacity to adjust smoothly. He concludes that 'The crash that will follow must make it clear that industrial society as such—and not just its separate institutions—has outgrown the range of its effectiveness.'[8]

We now go on to examine the issues raised in this introductory section. What are the forces which are thought to shape societies; what futures are projected or foreseen for ours; is a post-industrial society evolving; what form would it take; will continued economic growth be a part of it; what will be the position of technology—a tyranny of high technology or a humanised intermediate technology; and what path will we take to future society—can we build on ideas and institutions already in existence, however radical, or will there need to be a catastrophe?

7.1   The Causes of Social Development

There is no definitive theory which explains the process by which societies change but, particularly in a period when there is an awareness of a potential for major changes, we are driven to search for some insights. Daniel Bell, for example, is not prepared to settle for the position where everything dissolves into interacting forces. He prefers approaches which attempt to identify the central forces shaping a society:

> For Max Weber, the process of rationalisation is an axial principle for un-
> derstanding the transformation of the Western world from a traditional to
> a modern society: rational accounting, rational technology, the
> rationalistic economic ethic and the rationalisation of the conduct of life.
> For Marx, the production of commodities is the axial principle of
> capitalism, as the business firm is its axial structure; and for Raymond
> Aron machine technology is the axial principle of industrial society and the
> factory its axial structure.[9]

This section examines the role of ideas and technology in moulding the character
of society. In the realm of ideas the part played by political–economic theories
has been chosen as the main example because they are self-interpretations of the
societies from which they emerged and reflect prevailing values and attitudes.

7.1.1 *Ideas*

The term is used here in the broadest sense to represent man's values and his
explanations and interpretations of his situation in the world. These are a
major influence on his behaviour and a significant factor in shaping the
character of society. So, in the context of present-day environmental concerns, one might attempt to trace the origins of these problems in the world
view which has influenced the development of our society.

What have been the main features of the Western world view? John Black
suggests the following: (a) the belief that man's role in the world is to exploit
nature to his advantage, (b) the expectation that the human population will
expand, (c) a linear concept of time and a belief in progress, (d) a concern for
posterity.[10]

The most important source of these ideas, particularly in the case of attitudes to the natural environment, says Black, was Hebrew theology and its
Christian interpretation. Genesis, chapter 1, says 'Be fruitful and multiply,
and replenish the Earth and subdue it; and have dominion over the fish of the
sea and over the fowl of the air and over every living thing that moveth upon
the Earth'. Of course there is a potential conflict between the exploitation of
nature, if carried far enough, and the well-being of man. This conflict was
mediated in Hebrew and early Christian societies by a sense of responsibility
to the family (seen as extending into future generations) and to God—a duty
of stewardship.

Rights and responsibilities in the use of resources have been made concrete in the concept of property. This can vary widely from society to society, and the differing interpretations of personal property in the history of the West serve to illustrate the role of ideas in underpinning a social system, and the responsiveness of ideas to changing conditions. For example, John Black comments on the tendency for the balance between the property owner's rights and the wider interest to shift in favour of the individual at the time when the capitalist system was developing, and he quotes Locke's contemporary justification of unequal and unlimited capital accumulation.[11] This idea of unhindered individual decision on the allocation of resources (by producers and consumers) was central to the development of capitalism. But, an equally central feature of today's analysis of environmental problems is the inadequacy of counterbalancing mechanisms to represent collective interests. For this reason it is important to examine the economic and associated political ideas which have supported the development of the modern capitalist system.

Imagining the subsistence economy of a pre-commercial and industrial society, the issues which dominated the minds of classical and later economists would not have arisen. What was produced, and how much, would be determined by the farmer's own choice and the limits set by nature. The income of a family would simply be the result of its own efforts and, with self-sufficiency, the question of the market value of a product would not arise.

The coming of industrial society, characterised by the separation of production from consumption; the ability to tap massive energy reserves; the division of labour into specialised occupations and the establishment of a well-defined organising role, raised new issues. For example, how would it now be decided when enough had been produced, or which products to make. What was the basis of a product's market value—was there an objective standard such as the amount of labour involved in it, or was it determined by the individual user's valuation? If a labour theory of value was accepted, how could the surplus accumulated by the capitalist be justified? Classical economic theory, developing alongside the industrial system from the late eighteenth century, wrestled with these problems reflecting the concerns of the day. Light was thrown on the question of how much to produce by the formation of the idea that the need for products was unlimited, and that there was even a psychological need for the process of acquisition itself. Thus continuous economic growth was justified. It was held that the individual ascribed value to a product as measured by how much he paid for it. Labour was not regarded as the fountain of all value, the organising role of the *entrepreneur* was also a major contributor. It was said that the relative values of their respective contributions and the distribution of incomes, was determined by the impartial hand of market forces. Thus the accumulation of wealth was justified.

Weisskopf describes the idea underlying this economic theory as 'value-relativism', and concludes that 'The free market philosophy has contributed its fair share to the destruction of social and individual morality'.[12] All products which could be marketed profitably were good, irrespective of what they were or who got them. There was no objective criterion which could be applied to determine that one product was intrinsically more valuable than another, or that a given product would yield more satisfaction in different hands. There was thus no basis for interfering with the market process, which was seen merely as a neutral mechanism co-ordinating the rational decision-making of millions of individuals. This value-relativism had similar implications for the political system. It could be regarded as a market in which conflicts between different group interests were played out. In both cases, the hidden hand of the market process resolved the conflict and resulted in the greatest good of the greatest number. There was thus no basis for intervention, no other objective test of the common good.

It has long been recognised that these automatic solutions are not complete. For example, there are untidy residuals such as pollution and resource depletion where all the relevant information has not been processed. Later economic theories were thus able to justify intervention in order to refine market mechanisms without changing the underlying market idea. However, in Western societies today, the presence of giant, often multinational corporations and extensive government bureaucracies make the hidden hand more visible. To say the least, large business firms are not subject to the will of the consumer in quite the way that classical economic theory supposed. In the same way, effective participation in politics is more difficult in mass societies where the issues for decision are increasingly technical, complex and international in scope.

With the benefit of hindsight we may have acquired acceptable accounts of the process of change in ideas and society in the past, and gained some insight into the causes of contemporary problems. But what of the future? Jacques Ellul describes the insistence on rationalising all human activity as the most worrying form of determinism in the modern world.[13] This process of rationalisation, illustrated by economic theory and the development of large-scale industrial and governmental organisations, is the idea which Max Weber saw as the key to explaining the modern world. Its influence can also be traced through the development of science and technology. A world view which sets man apart from nature and urges him to dominate it, supports the idea of scientific enquiry. Schwartz sees a transition via science from a subjective to an objective, ahuman and, in the end, inhuman approach. Man himself becomes an object in a mechanical world, to be studied and controlled: 'Science has become a secular religion; technology its temple, efficiency its dogma'.[14]

7.1.2 *Technology*

Different writers analysing the impact of technology on society have offered their own interpretations of the key elements in the link between the character of a society and its technology. Some have stressed the dominant material or source of energy in use, others the impact of particular techniques either in the form of mechanical, social or intellectual technologies or even the mode of communication.

An example of the belief that the key to a phase of technology was its materials and its energy source, which have both environmental and cultural implications, is provided by Lewis Mumford. He coined the terms 'eotechnic' and 'paleotechnic' to describe the eras of wind and water and of coal, and applied the label neotechnic to the emerging twentieth-century technology.[15] Taking his paleotechnic period as an example, the physical environment would clearly be changed but so would the whole social system. By contrast with traditional pastoral or agricultural economies, the link can readily be seen between the large-scale use of fossil fuel and the use of machinery, the factory system, urbanisation, and the role of capital accumulation in the production process and, along with these, change in the whole structure of social relationships.

Alternatively, one could focus on the role of a particular technique in shaping a society. The motor-car, for example, has had a significant effect on our way of life, for exceeding the modest promise, or threat, of a mere horseless carriage. Many other examples can be thought of where particular techniques have far-reaching implications for present and/or future society. In the realm of social technology there is the giant corporation and also modern government with its military, administrative and welfare functions. In the case of technology more conventionally defined, three major examples are modern medicine, nuclear energy (either civil or military) and computers.

One aspect of technology which has received much attention in recent years is the communications media. For Harold Innis, and later Marshall McLuhan, communication is the most critical factor, not so much what is communicated but how the media operate. Innis attempted to analyse the development and effects of different modes of communication through history. He distinguished between oral and written forms; writing and printing; the different materials used, and between modes of transmission. A society's dominant form of communication imparts a bias.[16] For example, the use of tablets made from scarce and heavy clay gave early Middle Eastern civilisations a 'temporal bias'. Their geographical horizons were limited and their attention focused more on the flow of time: on law, tradition, religion. Papyrus, on the other hand, being abundant and readily transported, gave a spatial bias to Roman civilisation. It is not necessary to accept the historical generalisation of Innis to appreciate his point—that the character of the medium of communication will exert some influence on the nature of a society. Imagine the development of Western civilisation without writing. Think of modern society without the electronic transmission of information. What implications would follow if all communications had to be verbal and face to face? Less fancifully, consider Innis's point that the medium of communication influences power and politics in a society. It affects the amount and type of information which can flow and also, on the question of who is to control it, favours particular groups. This applies not only to the press, radio and television, but also access to information relating to the administration of business and governmental affairs.

An obvious comment on the approach which centres on the impact of one technique is that other forces must also have been involved in defining a society. This is the position taken by Jacques Ellul. He extends the concept of technique beyond the mechanical and electronic to the social, economic, administrative and psychological spheres—to include business corporations and social welfare organisations, political and governmental systems and techniques of persuasion and propaganda. For Ellul, it is this elaborate web of interlocking techniques which characterises our societies. Without apparent human intervention the web grows, divides and becomes universal. Technical problems become more complex, choices become more automatic and technicians come to dominate the management process.

To focus on the role of industrialism in shaping a culture is, in effect, to take this same approach. Industrial society is a collective term which can be used to describe the dynamic system of interrelationships between techniques, institutions and social relations which are of conern to Ellul. This logic of industrialisation transcends capitalism, but one of the most prolific and widely-read writers on the subject, J. K. Galbraith, has written in the context of Western societies. In contrast to the competitive, decentralised market model, Galbraith's books have charted the development of the new industrial state. This is a managed capitalism. Taking the U.S.A. as his example, Galbraith maintains that the economy is dominated by a few hundred mature corporations in which decision-making, rather than being a function of a separate, traditional management, is embedded within the organisation.[18]

He argues that the main goal of the technical-managerial elite, the technostructure, is its own survival and decision-making autonomy. It is in the interest of this control as well as the elite's income and prestige, that the corporations grow by expansion and merger—increasingly without regard for national boundaries. The government is a major buyer of the corporations' products (particularly of armaments and space equipment) and has also undertaken to maintain a high level of general demand in the economy through its management of fiscal and monetary policies. Furthermore, the government finances research, sanctions, the private enterprise system and the autonomy of the technostructure, and acts as a lender of last resort if things go wrong. Galbraith uses the term technocomplex to describe this symbiotic relationship between industry and government.

Not surprisingly, even such a brief examination of the forces which shape a society seems to point to a concept of social ecology which says that everything is related to everything! Social systems are constantly in motion as each element reacts to changes in the others. This applies to people as well as institutions and techniques. Our livelihood and security are bound up in present society and this is a strong influence on our responses and actions. Our characteristic collective response to problems, says Illich, has been escalation—more output, more science and technology, more police, schools, medicine, etc.[19] But these actions, which he would regard as self-defeating, are not the only form taken by our adaptation. Although it may be very difficult, impossible even, for us to see it actually happening to ourselves, might not our adaptation also take the form of changed perceptions, ways of thinking and values? Thus, taking the example of technology, Mumford feels that the machine, although it is a reflection of one side of his personality, has created an environment which now moulds man.[20] It has induced him to deny other elements of his make-up in favour of the rationalising, institutionalising demands of technology.

7.2 Thinking about the Future

We have no adequate explanation of the process of social change, and so do not know whether man is capable of imposing his will on the development of society, or if he is subject to blind forces such as technological determinism. In an era of very rapid change, and with specific problems thought to be looming, this becomes less of an academic question. There is an acute interest in the future. This section briefly contrasts two approaches to thinking about the future—one is neutral on the question of man's role, the other is involved.

The objective, neutral, scientific approach is illustrated by Herman Kahn and Anthony J. Wiener. They outline a 'basic, long-term multifold trend', express its components statistically where possible and project them into the future. This 'surprise-free' projection they see as providing a picture of a 'standard world' to which they add several 'canonical variations'. They illustrate the approach via futurology which Daniel Bell describes in his introduction to their book as a study of the future which is oriented to social policy. The aim of its methodology is to provide information and to sketch alternative futures upon which choices can be based. This does not involve prophesy but the identification of the constraints on our choice—resources, custom, will and basic trends in human society such as the growth of science, literacy and economic interdependence.[21]

In such a neutral extrapolation of trends, human perceptions and values could be regarded as just one of several mutually dependent variables. Each may change, and no special significance need be attached to the human element. After all, it might be argued, there is no reason for supposing that one state of human society—in the past, now, or in the future—is any more, or less natural than another. What is the essence of humanness in any case, it might be asked? Is it not enough to point out the interrelatedness of human culture, technology, etc. and project observable trends?

At the other extreme, for a writer representing the subjective, involved, humane approach, the human element is much more of an independent variable. He would probably establish an explicit set of values, a personal position—perhaps supported by evidence that certain conditions are essential for human well-being, and point to symptoms of malaise arising from the clash between this ideal and contemporary reality. Roger Williams quoted several authors' visions of the future which illustrate this involved approach.[22]

The emerging technological Utopia will be 'the deadliest, most sterile passivity history has ever known'; 'a comfortable, smooth, reasonable, democratic unfreedom'; 'an ice-age of perfect social functioning—a Utopia of changelessness.' The implication is that present trends are towards an homogenised, manipulated, dehumanised society. An illustration of this position explicitly stated is provided by Weisskopf. His focus is not on the impact of specific technologies but the influence of the sort of rationalism on which science and technology is based. He concludes that 'Western civilisation suffers from a gigantic repression of important dimensions of human existence', because of the tendency to reduce reason to that which is instrumental, technical, formalised, and to eliminate values, morality and ethics as dimensions of life and thought.[23] Thus, he illustrates the approach that takes man as an independent variable, and proposes that change be sought in order to match future society with what is believed to be the essential and whole human character.

Such a change is called for by Illich in his book *Tools for Conviviality* (1973). The tendency of modern productive tools, machines and institutions is to deny to the individual the freedom to use them when he wishes for purposes determined by him. Illich urges that the concept of convivial tools, which 'enlarge the range of each person's competence, control and initiative', be used as a guideline in the structuring of future society. The existing tendency towards 'specialisation of functions, institutionalisation of values and centralisation of power' is turning people into the 'accessories of bureaucracies or machines.' His fundamental point is that there is a limit to this process:

> ... only within limits can machines take the place of slaves; beyond these limits they lead to a new kind of serfdom. Only within these limits can education fit people into a man-made environment: beyond these limits lies the universal schoolhouse, hospital ward or prison. Only within limits ought politics to be concerned with the distribution of maximum industrial outputs, rather than with equal inputs of either energy or information.[24]

Illich typifies what we have called the involved, humane approach. When looking at man and his future he does not adopt the neutral, value-free method of the systems analyst, which incidentally he would see as a manifestation of the impact of the industrial system on human thought and values. Instead, he regards man, his values and goals, as a limiting factor—an independent variable. Man can adapt, but only within limits, whereas the present industrial system is dynamically unstable.[25] Thus the scene is set for catastrophe as the industrial system pushes inexorably towards these limits. If this is to be avoided, these human limits must be explored in order to guide us as we make those succeeding choices which will eventually determine the shape of our future.

7.3   Technology, Growth and Change

Are we evolving towards a post-industrial society? The concluding section of this chapter approaches this question by focusing on three aspects of future society: the role of technology; the implications of continued economic growth and the concept of a 'steady-state' economy; and the possible paths to the future.

7.3.1   *Tyrannical or Humanised Technology?*

In the same way that some writers have projected current trends in the use of resources to produce an arresting vision of a depleted world, others have tried to foresee the development of society. One major determinant of the character of future society will be the role of technology. So powerful is the combination of social, political, economic and cultural forces which seem to have created, and which reinforce our technological society, that a strong current of technological determinism runs through the literature. This takes the form of cautionary analyses of technology's historical impact and its current trends, or might even be said to be illustrated directly by those who have faith in continuing technological solutions to contemporary problems. The critics, pointing to a future society tyrannised by technology (which demands that the environment, and man, be adapted to fit), call for a humanised technology.

Earlier sections of this chapter have shown that technology does not just mean machines—although particular devices can have great significance for our lives. The term can be used as in social technology (organisations) and intellectual technology (systems analysis, for example). In fact, although difficult to define precisely, technology also derives its impact from the rationale which underlies it—control, optimisation, efficiency. The fear of many writers is that these values, and the systems embodying them, have gone too far. Technical change, intimately bound up with the pursuit of economic growth, has created a 'gale of creative destruction' and contributed to the alienation, insecurity, rootlessness and lack of community which threatens the physical and mental stability of society.[26] The search for efficiency has led to the creation of massive organisations and has concentrated power of control in few hands. These same values have also influenced governments. Now, says Ellul, Democracy becomes an illusion fabricated for the purpose of ensuring public apathy, 'man, practically speaking, no longer possesses any means of bringing action to bear on technique'.[27]

It is the consolidation of such a 'technocratic totalitarianism' which is feared by Theodore Roszak and depicted in his *The Making of a Counter Culture* (1969). With its 'techniques of inner manipulation as unobstrusively fine as gossamer', it will adapt to us to 'an existence wholly estranged from everything that has made the life of man an interesting adventure'.

> Above all, the capacity of our emerging technocratic paradise to denature the imagination by appropriating to itself the whole meaning of Reason, Reality, Progress, and Knowledge will render it impossible for men to give any name to their bothersomely unfulfilled potentialities but that of madness. And for such madness, humanitarian therapies will be generously provided. The question therefore arises: If the technocracy in its grand procession through history is indeed pursuing to the satisfaction of so

many such universally ratified values as The Quest for Truth, The Conquest of Nature, The Abundant Society, The Creative Leisure, The Well-adjusted Life, why not settle back and enjoy the trip? The answer is, I guess, that I find myself unable to see anything at the end of the road we are following with such self-assured momentum but Samuel Beckett's two sad tramps forever waiting under that wilted tree for their lives to begin. Except that I think the tree isn't going to be real, but a plastic counterfeit. In fact, even the tramps may turn out to be automatons ... though of course there will be great programmed grins on their faces.[28]

Ivan Illich provides a good example of those making a plea for the humanisation of technology.[29] He speaks of natural scales in several dimensions of human life. The growth of organisations beyond these points is a threat to society, isolating men and undermining the texture of community 'by promoting extreme social polarisation and splintering specialisation.' This represents a 'crisis in the industrial mode of production itself.' What is needed is the enlargement of 'the range of each person's competence, control and initiative, limited only by other individual's claims to an equal range of power and freedom'. Tools (machines, institutions, systems) which facilitate this he calls convivial. Industrial tools are manipulative and 'allow their designers to determine the meaning and expectations of others'.

The character of a tool is not simply a function of the level of technology. Convivial does not necessarily mean small scale. The telephone system 'lets anybody say what he wants to the person of his choice', whereas a hand-tool, the dentist's drill, 'can be restricted by becoming the monopoly of one profession'. Illich makes the point, therefore, that 'The criteria of conviviality are to be considered as guidelines to the continuous process by which a society's members defend their liberty, and not as a set of prescriptions which can be mechanically applied.' Some industrial modes of production are anti-convivial but their manipulative tendencies might be countered by institutional arrangements—promoting public accountability, industrial democracy or consumer power. 'It is a mistake to believe that all large tools and all centralised production would have to be excluded from a convivial society. It would equally be a mistake to demand that for the sake of conviviality the distribution of industrial goods and services be reduced to the minimum consistent with survival in order to protect the maximum equal right to self-determined participation.' It is a question of achieving a balance between 'distributive justice and participatory justice'.

Although this sounds superficially reasonable and moderate, in practice it has radical implications. It would mean the rejection of some technological the points about growth and human welfare, and the problems of physical from them) in favour of more convivial arrangements. Rather than specify all the obstacles which would present themselves, we might ask 'Which influential groups in society would currently support such changes—workers, employers, government?'

7.3.2   *Growth or a Steady-state Economy?*

From several viewpoints, economic growth could be regarded as the distinguishing feature of the modern world. The fact of revolutionary economic advance within the last century or two defines modern societies; the concept of growth has underpinned modern economic theory; the already industrialised nations have taken the long-term continuation of growth completely for granted and its promotion has been a keystone of government and corporate policies; the concept of development in the rest of the world has been conventionally defined in terms of economic growth.

Apparently a simple measure of an economy's output of goods and services, on examination the concept of economic growth turns out to be slippery. What is the measure to include or exclude? how is an item to be valued? what is the relationship between the level of economic output and human welfare? Typically, in a Western society, only those goods and services which enter the market are included, and they are valued by the market. Family or voluntary services performed outside the market are not included, and state services such as defence, administration, education and medicine, are included, but at cost because they are not directly marketed. Already, economic output as an accurate measure of human welfare is put in doubt. In addition, might some of the elements of output be regarded more as *costs* than benefits? For example, the exploitation of non-renewable resources, waste disposal, transportation, the treatment of industrial injuries, and some education and training? Also, the contribution of the economy to a community's welfare could be related not only to what things are produced but how they are shared between families. Finally, is not the standard of living, welfare, the quality of life also related to non-economic features of life?

The view that increasing material prosperity is not necessarily commensurate with improved human welfare has a long history. In more recent times, however, two other arguments have combined with it to undermine confidence in the role of continued economic growth. These arguments can be seen in chapter 4—that there are physical limits to growth in terms of resource depletion and waste disposal, and that there are increasing human costs of adaptation to the demands of technology and to the pace of change. One response to this position was seen in chapter 6. This accepts in principle options (together with the increased output and incomes which would follow and human limits. But, it would propose that, in practice, the link between growth and welfare *is* positive—more output, for example, means that poverty and pollution can be tackled. What is needed, the argument might go, is *better* growth. The mechanisms by which a society compares the costs and benefits of growth should be refined and alternative measures of welfare developed. Together with the application of new technology this would also lead to cleaner and more sustainable growth.

Alternatively, some people would not accept the accuracy of such judgements, because they take a longer view; because they come to different conclusions about the imminence of catastrophic resource depletion and environmental degradation, or because they regard such arguments as the special pleading of a society hooked on growth and technological solutions. Some writers thus advocate, or at least explore the concept of, a steady-state economy. By contrast with the conventional economy which aims to maximise throughput, the object would be to achieve a constant level of physical wealth with the minimum of throughput. The emphasis is on the stock rather than the flow of goods. With a constant population, which is usually also proposed, the amount of necessary production would depend on the level of physical wealth desired and the durability of its components. Of course, the environmental impact of man's economic activities would not be reduced to zero, even if levels of population and output were set much lower than now. Advocates of the steady state would argue, however, that at least the estimated time of arrival at various human and ecological frontiers would be delayed. Furthermore, changes of attitude might be prompted by the new experience which would improve our ability to meet such problems.

Certainly, compared with societies where, in Weisskopf's words, 'there is no rational basis or moral constraint on consumption and production, where consumption has become a metabolic process—ingest–digest–discharge,'[30] a steady-state economy would be associated with radically different attitudes. In effect this is to say that the transition to a steady state from a growth-orientated economy has enormous social, political and economic implications. Naturally, there is the problem of how to decide on an appropriate level of population and physical wealth, but also how are these to be controlled and how is the fixed stock of goods to be shared? In a growing economy the heat can be taken out of conflict over income differentials by offering more to each group in turn. The differentials may not be removed but each group gets progressively more in absolute terms. In the steady-state economy income distribution would become a zero-sum game, one group's gain could only be at the expense of others. Similarly, with a limit set on the amount of production, an increase in the welfare of the community as a whole could only be achieved by producing a better combination of goods or by redistributing it between groups of people in such a way as to raise the total satisfaction derived.

Conventional economic theory, unable to resolve these questions, left them to the hidden hand of the market. So, incomes were said to be determined by labour productivity. But if growth of output is no longer desired, and increased productivity means less input rather than more output, then productivity may be undermined as a basis for deciding income rewards. This would be especially true if restrictions on output meant much reduced levels of employment. Would the lottery of unemployment in a market economy be regarded as part of the process of income determination, or would incomes be decided on other than market considerations? If incomes were determined on criteria other than work contribution then the unemployed man would not necessarily receive less than the employed. This raises the question of the merit of unemployment (or employment). It might be decided that employment was inherently desirable, satisfying a social or psychological need. If so, labour saving methods of production would be less attractive, especially if their resource and energy demands were high. This amounts to matching technology to population (employment needs).

The question arises, how will these or any other changes come about? How would the decision to match technology to population be made and implemented? The same applies to the levels of population and production. 'Mutual coercion, mutually agreed upon,' says Garrett Hardin,[31] but what would this involve? Could output be controlled by rationing resources? Would a consumer nation limit its use of scarce resources unilaterally, leaving other nations free to exploit them? Will the countries with extractive and agricultural economies ration supplies? Will international co-operation and authority be needed? These questions about the process and political implications of change are taken up in the final section.

7.3.3 *Evolution or Catastrophe?*

Whatever the exact nature of the problems which may emerge in the next few decades, a central question for social scientists is: In what way will societies respond? In general terms the response will be shaped by the interaction between three variables: the distribution of effective decision-making power; the information which is being fed in; and the framework of values within which it is interpreted. These categories are by no means completely independent, but the examination of possible changes in them does shed light on the way societies might respond in the future.

Take the problem of resource depletion as an example, and assume no change in these variables. The structure of prices would alter, production and consumption would adapt, and technology would be applied to reduce resource requirements per unit of output, recycle wastes and find alternative materials. This process of adaptation would probably not be smooth and crises would occur from time to time, but more fundamental doubts have been expressed by writers who believe that this approach would not work in the end. Schwartz argues that a technological solution is never complete, each quasi-solution generates a residue of other techno–social problems, and in any case problems arise faster than solutions and are progressively more difficult and costly to handle.[32] Further more, it is just this projection of present trends which has led Ellul, Roszak and others to foresee their technological anti-Utopias.

Could existing values and decision-making processes cope if they had better information? One line of reform suggests that the responses of the industrial system to human and environmental problems would be improved if the market mechanism could be supplemented by information in the form of taxes, subsidies, legislation and social indicators reflecting previously hidden costs and benefits.

An alternative to this carrot-and-stick approach would be to restructure the decision-making process. Decisions previously made by the decentralised processes of the market might be communalised. New state institutions might ration resources and control the development of science and technology. Pirages and Ehrlich describe a Planning Branch of government with five sections:

(a) Environmental Protection—conducting environmental impact studies, making projections and outlining legislation.

(b) Natural Resources—evaluating reserves and needs, and planning rates of use.

(c) Social Ecology—monitoring changing social conditions and overseeing the social aspects of transition to a steady-state economy; recommending social legislation, education programmes and population policies.

(d) Economic Priorities—concerned with allocation of capital and income distribution.

(e) Technology Assessment—evaluating all major new techniques, outlining research and development priorities, investigating potential social–environmental–economic impact, devising new methods to meet energy and resource crises.[33]

In *The Coming of Post-industrial Society* (1973), Daniel Bell stresses this movement towards communal decision-making:

> The decisive social change taking place in our time—because of the interdependence of men and the aggregative character of economic actions, the rise of externalities and social costs, and the need to control the effects of technical change—is the subordination of the economic function to the political order.

He sees far-reaching implications in this change:

> We now move to a communal ethic, without that community being, as yet, wholly defined. In a sense, the movement away from governance by political economy to governance by political philosophy—for that is the meaning of the shift—is a turn to non-capitalist modes of social thought. And this is the long-run historical tendency in Western society.[34]

'The crucial problem for the communal society', Bell observes, 'is whether there is a common framework of values that can guide the setting of political policy'.[35] With decision-making processes more visible and concentrated, how will such a consensus be reached, even if effective forms of participation can be devised? If manipulation or repression is the response to the increased level of conflict which Bell concludes is likely, then the possibility of Roszak's technocratic totalitarianism remains.

These examples of the possible responses of social systems to future problems have assumed changes in decision-making processes or information, but no autonomous change in the framework of values. Given different values, then more fundamental reshaping of society can be envisaged. For example, assume that, with Illich, we all rejected 'the bureaucratic management of human survival' and accepted the values implicit in a convivial and steady-state society (see sections 7.3.1 and 7.3.2). The emphasis would then be on participative justice, optimising the individual's power of self-determination, with the level of output and its distribution being held constant by agreement. A society like this is possible in theory, but in practice it would be radically different from what we know. How could it come about? Those committed to bringing about these changes define strategies by reference to the cumulative actions of affinity groups (ecological action and anti-growth groups), radical caucuses (in political parties and professional associations) and counter institutions (free universities, communes, worker-owned firms).[36] Alternatively the change could be sudden:

> Some fortuitous coincidence will render publicly obvious the structural contradictions between stated purposes and effective results in our major institutions. People will suddenly find obvious what is now only evident to

a few: that the organisation of the entire economy towards the better life has become the major enemy of the *good* life. Like other widely shared insights, this one will have the potential of turning public imagination inside out. Large institutions can quite suddenly lose their respectability, their legitimacy and their reputation for serving the public good.[37]

Illich goes on to point out that there are historical precedents for changes of this sort. Also, although the present system of values may seem to be totally dominating and continuously reinforced by experience, Black reminds us that 'different models of the universe exist side by side and remain available, if only in skeletal forms, to be taken to pieces and used as building blocks for a new model.'[38]

Of course no one can predict just what problems will emerge, or how societies will adapt to them. It is probable that all the elements just described will be involved—a little bit more of the mixture as before: piecemeal reforms; the extension of state decision-making; and possibly a change in the framework of values, accompanied by more radical changes in the social system. The pace at which problems emerge, and their magnitude (the information we receive from the natural and human environment) will influence the speed and nature of social change. Catastrophe—a sudden, major change—represents one extreme. On the other hand, anticipated problems are also information and could evoke a more measured response. Which segment of the spectrum will apply to us?

One more complicating factor must be added. It is not enough to consider changes within societies. The issues with which this book has been concerned have clear international dimensions. The impact of some pollutants is global. The Earth's resources can effectively be regarded as a common property resource. Is a single nation likely to limit its use of resources, leaving them to be depleted by those who do not? Would one military giant forego economic growth and high technology if the other did not? Given such a variety of cultures, ideologies and levels of development, can all the nations of the world be expected to have the same schedule of preferences with regard to population, economic growth, resource use and pollution?

Conclusion

The first four chapters of this book have presented the background to topics such as ecology, resources, population and pollution, and traced the development of the debate which together make up the environment issue. The succeeding chapters, in order to provide a basis for understanding and evaluating the policy approaches which might be adopted, and to put the different positions in the debate into perspective, have analysed environmental issues in their social, economic and political contexts. Essentially, these chapters see environmental problems as the outcome of complex processes of choice in society. Decision-making procedures have been outlined and examples given of resource misallocation and their possible causes. The obvious implication of this analysis is that changes could be made to correct defects in the relevant decision-making processes. Chapter 6 has illustrated the possible roles of taxes, subsidies, regulations, etc. in dealing with pollution and the even more complex and difficult problems presented by resource depletion and developments in science and technology.

One of the most important points we have sought to make is that, initially at least, environmental problems can usefully be seen, not as acts of God or indeed in any special category, but rather as falling within the range of everyday political–economic issues. The approach to the analysis and treatment of environmental problems in chapters 5 and 6 is based on this premise, and a wide range of policy options can be based on it. However, it is arguable that some environmental problems do have special characteristics. The introduction to chapter 7, for example, poses the question: Could it be that some combination of the scale of environmental problems, the pace at which they appear and the way they combine with each other, demands changes which go beyond the framework of existing society? Would the reformist approach outlined in chapter 6 prove to be adequate?

The evidence of chapter 7 is that more radical changes could well be needed, and of course the broad pattern of history is composed of such paradigm shifts. But what shape would such changes take, and how would they come about? Chapter 7 has approached these questions by examining some of the active ingredients in past changes (ideas and technology) and the methods of analysis and views of some contemporary writers who have turned their attention to the possible character of future society. The purpose of this chapter is to complete the analysis of chapters 5 and 6 by indicating the potential for the more fundamental changes in social, economic and political systems which the solution of some environmental problems may demand, and which history confirms do happen.

Inevitably in a book which aims to provide an introduction to the analysis of environmental problems rather than to state a position in the environment debate, the concluding chapter ends with a question mark. In particular, the final paragraph echoes the image of the spaceship Earth presented in the first sentence of chapter 1. This global perspective is a major feature of the multifaceted issue which we have come to know simply as *the environment*. Within nations, the technical, economic and political demands of effective environmental policies will continue to provide a challenge to the natural and social sciences alike. But perhaps the most intriguing question is how, if at all, the community of nations will respond to that image of a spaceship Earth given to us by the first astronauts and which is made manifest in global aspects of the environment issue such as pollution and the sharing and depletion of resources.

References

1 Basic Ecological Principles

1. Report of the Hydrological Decade Conference, 1970

2. B. Bolin, (1970). 'The carbon cycle', *Scientific American*, **223,** no. 3

2 Population and Food

1. D. J. Bogue (1969). *Principles of Demography* (Wiley, New York and London)

2. United Nations (1968). *World Population Prospects* (U.N., New York)

3. Ibid.; and J. Durrand (1971). 'The Modern Expansion of World Population', in *Man's Impact on Environment* (ed. T. Detwyler) (McGraw-Hill, New York and London)

4. United Nations (1968), op. cit.

5. L. Brown (1970). 'Human Food Production as a Process in the Biosphere', in *Scientific American* symposium *The Biosphere* (W. H. Freeman, San Francisco) p. 165

6. G. Bergstrom, quoted in P. and A. Ehrlich (1972). *Population, Resources, Environment* (W. H. Freeman, San Francisco)

3 Energy, Materials and Pollution

1. U. S. Bureau of Mines (1970). *Mineral Facts and Problems* (Government Printing Office, Washington, D. C.)

2. K. Irikayama (1967). 'The Pollution of Minimata Bay and Minimata Disease', *Advances in Water Pollution Research*, 3

3. N. E. Cooke (1971). Symposium report on Mercury in Man's Environment, *Environment*, **13,** no. 4

4. T. Aaronson (1971). 'Mercury in the Environment', *Environment*, **13,** no. 4

5. S. C. Gilfillan (1965) 'Lead Poisoning and the Fall of Rome', quoted in P. and A. Ehrlich (1970). op. cit., p. 166

6. N. Blumer (1970). Scientific Aspects of the Oil Spill Problem, a paper presented at the Oil Spills Conference (Committee on Challenges of Modern Society, NATO, Brussels)

4 The Environmental Debate

1. B. Commoner (1963). *The Closing Circle* (rev. edn. 1971, Cape, London) p. 186

2. P. and A. Ehrlich (1972). op cit., pp. 432–43

3. D. Meadows *et al.* (1972). Club of Rome report, *The Limits to Growth* (Earth Island, New York) pp. 153–154

4. E. Goldsmith *et al.* (1972). *A Blueprint for Survival* (Penguin, Harmondsworth: book publication of *Ecologist* articles of January 1972) pp. 30–31

5. E. G. Dolan (1971). *TANSTAAFL: The Economic Strategy for Environmental Crisis* (Holt, Rinehart & Winston, New York) pp. 11–12

6. D. Bell (1973). *The Coming of Post-industrial Society* (Basic Books, New York) p. 480

7. W. A. Weisskopf (1971). *Alienation and Economics* (Dutton, New York) p. 52.

8. S. H. Nasr (1968). *The Encounter of Man and Nature* (Allen & Unwin, London); quoted in John Black (1970). *The Dominion of Man: The Search for Ecological Responsibility* (Edinburgh University Press) pp. 19–20

9. J. Black (1970). op. cit., p. 121

10. L. White, Jr (1971). 'The Historical Roots of Our Ecological Crisis', quoted in *Man's Impact Upon Environment*, (ed. T. Detwyler). op. cit.

11. E. F. Schumacher (1973). 'Modern Pressures and the Environment', *Christian Action Journal*, summer 1973

12. J. V. Taylor (1975). *Enough is Enough* (SCM Press, London)

13. W. Kuhns (1971). *The Post-industrial Prophets: Interpretations of Technology* (Weybright & Talley, New York) p. 89

14. E. Shils; quoted in François Hetman (1973). *Society and the Assessment of Technology* (OECD, Paris) p. 35

15. A. Chisholm (1972). *Philosophers of the Earth* (E. P. Dutton, New York) pp. 131–132

16. H. Brooks (1967). 'Applied Science and Technological Progress', *Science*, **156**, p. 1712

17. E. S. Schwartz (1971). *Overskill: The Decline of Technology in Modern Civilisation* (Quadrangle Books, Chicago) p. *ix*

18. F. Hetman (1972). op. cit., p. 37

19. W. Kuhns (1971). op. cit., pp. 60–61

20. A. Chisholm (1972). op. cit., p. 19

21. E. Boulding (1971). The Economics of the Coming Spaceship Earth, in *The Environmental Handbook* (ed. John Barr) (Ballantine with The Friends of the Earth, London) pp. 77–82

22. E. J. Mishan (1967). *The Costs of Economic Growth* (Pelican, Harmondsworth) p. 64

23. H. V. Hodson (1972). *The Diseconomics of Growth* (Earth Island, New York) p. 238

24. W. A. Weisskopf (1971). op. cit., p. 92

25. Dolan (1971). op. cit., p. 12

26. I. D. Illich (1973). *Tools for Conviviality* (Calder & Boyars, London) p. 50

27. D. C. Pirages and P. R. Ehrlich (1974). *Ark II: Social Response to Environmental Imperatives* (W. H. Freeman, San Francisco) p. 64

28. L. K. Caldwell (1971). 'Environment and Administration: 'The Politics of Ecology', in *Environment, Resources, Pollution and Society* (ed. William W. Murdoch) (Sinauer Associates, Stamford, Connecticutt) p. 394

29. Dolan, (1971). op. cit., p. 14

30. D. C. Pirages and P. R. Ehrlich (1974). op. cit., p. 95

31. R. Dubos (1966). 'Promises and Hazards of Man's Adaptability', in *Environment Quality: Resources for the Future* (Johns Hopkins

University Press, Baltimore) pp. 38–9; reproduced in H. Jarrett (ed.) *Environmental Quality in a Growing Economy* (Johns Hopkins University Press, Baltimore, 1971)

32. I. D. Illich (1973). op. cit., pp. 12, 34

5 The Politics and Economics of the Environment

1. K. W. Kapp (1972). 'Environmental Disruption and Protection', a paper delivered at the Fourth International Conference of the German Metalworkers Union; translated in *Socialism and the Environment* (Spokesman Books, Nottingham, 1972)

2. A. G. Gruchy (1974). Government Intervention and the Social Control of Business, *Journal of Economic Issues*, **VIII,** no. 2

3. J. K. Galbraith (1952). *American Capitalism: The Concept of Countervailing Power* (Houghton Mifflin, Boston, Massachusetts; paperback edn, Penguin, Harmondsworth, 1963)

4. J. K. Galbraith (1967). *The New Industrial State* (Houghton Mifflin, Boston, Massachusetts, and Hamish Hamilton, London); and (1973), *Economics and the Public Purpose* (Houghton Mifflin, Boston, Massachusetts; Deutsch, London, 1974)

5. C. K. Wiber, (1974). 'Economics, Power and the Regulation of Multinational Corporations', *Journal of Economic Issues*, **VIII,** no. 2

6. Report of a conference on Science and Technology for Human Development, World Council of Churches, Bucharest, 1974; published in *Anticipation*, no. 9 (November 1974)

7. Mahbub ul Haq (1975). 'Toward a new Framework for International Resource Transfers', *Finance and Development*, **XII**, no. 3 (September 1975)

6 Environmental Policy

1. L. E. Ruff (1970). 'The Economic Common Sense of Pollution', *Public Interest*, 19, p. 72

2. R. C. D'Arge, and J. E. Wilen (1974). Governmental Control of Externalities, *Journal of Economic Issues*, **VIII,** no. 2

3. D. A. Bigham, (ed.) (1973). *The Law and Administration Relating to Protection of the Environment* (Oyez Publishing, London) p. 1

4. Ibid., p. *v*

5. Ibid., pp. 121–122

6. A. D. McKnight (1974). Law and Administration, in *Environmental Pollution Control: Technical, Economic and Legal Aspects* (ed. Allan D. McKnight, Pauline K. Marstrand, T. Craig Sinclair) (Allen and Unwin, London), p. 34

7. W. C. Osborn (1974). in McKnight, Marstrand, Sinclair (eds) op. cit., p. 274

8. D. Harris (1974). op. cit., p. 84

9. A. D. McKnight, P. K. Marstrand and T. C. Sinclair (eds) (1974). op. cit.

10. D. T. Savage, M. Burke, J. D. Coupe, T. D. Duchesneau, D. F. Wihry, J. A. Wilson (1974). *The Economics of Environmental Improvement* (Houghton Mifflin, Boston, Massachusetts) p. 168

11. P. Connelly and R. Perlman (1975). *The Politics of Scarcity: Resource Conflicts in International Relations* (Oxford University Press, London and New York) pp. 143–4

12. F. Hetman (1973). op. cit., pp. 340–341

13. Ibid., pp. 354–355

14. D. Bell (1973). op. cit., p. 335

15. Ibid., p. 326

16. Confederation of British Industry (1973). *A New Look at the Responsibilities of the British Public Company* (interim report of the committee under Lord Wilkinson)

17. J. Hamble (1973). *Social Responsibility Audit* (Foundation for Business Responsibilities; London)

18. Social Audit (1974). *Report on Tube Investments*

19. A. V. Kneese (1967). Water Quality Management by Regional Authorities in the Ruhr Area, in *Controlling Pollution: The Economics of a Cleaner America* (ed. Marshall I. Goldman) (Prentice–Hall, Englewood Cliffs, New Jersey) pp. 115, 119–21

20. Ibid., pp. 121–122

21. Ibid., pp. 126–127, and 122

7 An Evolving Post-industrial Society?

1. W. A. Weisskopf (1971). op. cit., p. 16

2. Sir Peter Medawar; quoted in F. Hetman (1973). op. cit., p. 37

3. A. Toffler (1970). *Future Shock* (Bodley Head, London)

4. R. Carson (1963). *Silent Spring* (rev. edn 1970, Fawcett, New York, and Penguin, Harmondsworth)

5. I. D. Illich (1973). op. cit., p. 104

6. A. Chisholm (1972). op. cit., p. 19

7. Ibid., p. 21

8. I. D. Illich (1973). op. cit., pp. 104, 108

9. D. Bell. (1973). op. cit., pp. 10–11

10. J. Black (1970). op. cit., p. 21

11. Ibid., p. 70

12. W. A. Weisskopf (1971). op. cit., p. 116

13. J. Ellul (1965). *The Technological Society* (Cape, London) p. *xxxi*

14. E. S. Schwartz (1971). op. cit., p. 4

15. W. Kuhns (1971). op. cit., p. 38

16. Ibid., p. 141

17. Ibid., pp. 88, 89

18. J. K. Galbraith (1973). op. cit.

19. I. D. Illich (1973). op. cit., ch. 1

20. W. Kuhns (1971). op. cit., p. 60

21. H. Kahn, and A. J. Wiener (1967). *The Year 2000: A Framework for Speculation on the Next Thirty-Three Years* (Macmillan, New York; Collier–Macmillan, London, 1968), pp. 7 and *xxviii*

22. R. Williams (1971). *Politics and Technology* (Macmillan, London), pp. 32–34

23. W. A. Weisskopf (1971). op. cit., pp. 16, 36

24. I. D. Illich (1973). op. cit., p. *xii*

25. Ibid., p. 46

26. W. A. Weisskopf (1971). op. cit., p. 170

27. W. Kuhns (1971). op. cit., p. 96

28. Theodore Roszak (1969). *The Making of a Counter Culture* (Double-day, New York; Faber & Faber, London, 1970) pp. *xiii, xiv*

29. I. D. Illich (1973). op. cit., pp. *xi, xii*, 20–24

30. W. A. Weisskopf (1971). op. cit., p. 116

31. G. Hardin (1973). 'Tragedy of the Commons', in *Towards a Steady State Economy* (ed. Herman E. Daley) (W. H. Freeman, San Francisco)

32. E. S. Schwartz (1971). op. cit., p. 76

33. D. C. Pirages and P. R. Ehrlich (1974). op. cit., pp. 176–177

34. D. Bell (1973). op. cit., pp. 373, 298

35. Ibid., p. 482

36. G. Lakey (1973). *Strategy for a Living Revolution* (W. H. Freeman, San Francisco) pp. 197–199

37. I. D. Illich (1973). op. cit., p. 103

38. J. Black (1970). op. cit., p. 119

Further Reading

1 Basic Ecological Principles

A. Boughey (1971). *Fundamental Ecology* (Intertext, New York and Leighton Buzzard, Bedfordshire)

E. J. Kermondy (1969). *Concepts of Biology* (Prentice-Hall, Englewood Cliffs, New Jersey and London)

Scientific American symposium (1970). *The Biosphere* (W. H. Freeman, San Francisco)

2 Population and Food

G. Borgstrom (1973). *World Food Resources* (Intertext, New York and Leighton Buzzard, Bedfordshire)

P. R. Ehrlich (1971). *The Population Bomb* (Ballantine Books, New York and London)

P. Ehrlich and A. Ehrlich (1970). *Population, Resources, Environment* (W. H. Freeman, San Francisco)

K. C. W. Kammeyer (1972). *An Introduction to Population* (Intertext, New York and Leighton Buzzard, Bedfordshire)

M. E. Solomon (1969). *Population Dynamics* (Edward Arnold, London; St Martin's Press, New York)

3 Energy, Materials and Pollution

R. Carson (1963). *Silent Spring* (Houghton Mifflin, Boston, Massachusetts, and Hamish Hamilton, London; rev. edn 1970, Fawcett, New York, and Penguin, Harmondsworth)

P. Cloud, (ed.) (1969). *Resources and Man* (W. H. Freeman, San Francisco)

J. Holdren and P. Herera (1972). *Energy* (Sierra Club Books, New York)

A. Tucker (1972). *The Toxic Metals* (Ballantine Books, London; Earth Island, New York)

K. Warren (1973). *Mineral Resources* (David & Charles, Newton Abbot; Halsted Press, New York; paperback edn, Penguin, Harmondsworth)

4 The Environmental Debate

A. Chisholm (1972). *Philosophers of the Earth* (E. P. Dutton, New York; Sidgwick & Jackson, London)

5 The Politics and Economics of the Environment

P. W. Barkley and D. W. Seckler (1972). *Economic Growth and Environmental Decay: The Solution Becomes the Problem* (Harcourt Brace Jovanovich, New York)

6 Environmental Policy

J. H. Dales (1968). *Pollution, Property and Prices* (University of Toronto Press)

E. G. Dolan (1971). *TANSTAAFL: The Economic strategy for Environmental Crisis* (Holt, Rinehart & Winston, New York and London)

7 An Evolving Post-industrial Society?

D. C. Pirages and P. R. Ehrlich (1974). *Ark II: Social Response to Environmental Imperatives* (W. H. Freeman, San Francisco)

R. Williams (1971). *Politics and Technology* (Macmillan, London)

Index